Plant Transformation via *Agrobacterium tumefaciens*

This book presents information from authentic and highly regarded sources in plant biotechnology. It provides a comprehensive review of plant transformation in soybean. The book discusses protocol inefficiencies, the type and conditions of explants, culture conditions, complexity in transformation processes, host-specificity and *Agrobacterium*-genotype interactions. The book provides insights promoting soybean use in the manufacturing of health-promoting supplements, feed production, food products and biodiesel production. It helps the reader in achieving successful soybean transformation and outlining the possibilities that could help reduce recalcitrance in soybean.

Plant Transformation via *Agrobacterium tumefaciens*

Culture Conditions, Recalcitrance and Advances in Soybean

Phetole Mangena

CRC Press is an imprint of the
Taylor & Francis Group, an **informa** business

First edition published 2023
by CRC Press
6000 Broken Sound Parkway NW, Suite 300, Boca Raton, FL 33487-2742

and by CRC Press
4 Park Square, Milton Park, Abingdon, Oxon, OX14 4RN

CRC Press is an imprint of Taylor & Francis Group, LLC

Library of Congress Cataloging-in-Publication Data

Names: Mangena, Phetole, author.
Title: Plant transformation via agrobacterium tumefaciens : culture conditions, recalcitrance and advances in soybean / Phetole Mangena.
Other titles: Culture conditions, recalcitrance and advances in soybean
Description: First edition. | Boca Raton, FL : CRC Press, 2023. | Includes bibliographical references and index.
Identifiers: LCCN 2022009578 (print) | LCCN 2022009579 (ebook) | ISBN 9781032250373 (hardback) | ISBN 9781032250380 (paperback) | ISBN 9781003281245 (ebook)
Subjects: LCSH: Agrobacterium tumefaciens. | Plant biotechnology. | Soybean.
Classification: LCC QR82.R45 M36 2023 (print) | LCC QR82.R45 (ebook) | DDC 577.5/7--dc23/eng/20220330
LC record available at https://lccn.loc.gov/2022009578
LC ebook record available at https://lccn.loc.gov/2022009579

ISBN: 9781032250373 (hbk)
ISBN: 9781032250380 (pbk)
ISBN: 9781003281245 (ebk)

DOI: 10.1201/b22829

Typeset in Times
by Deanta Global Publishing Services, Chennai, India

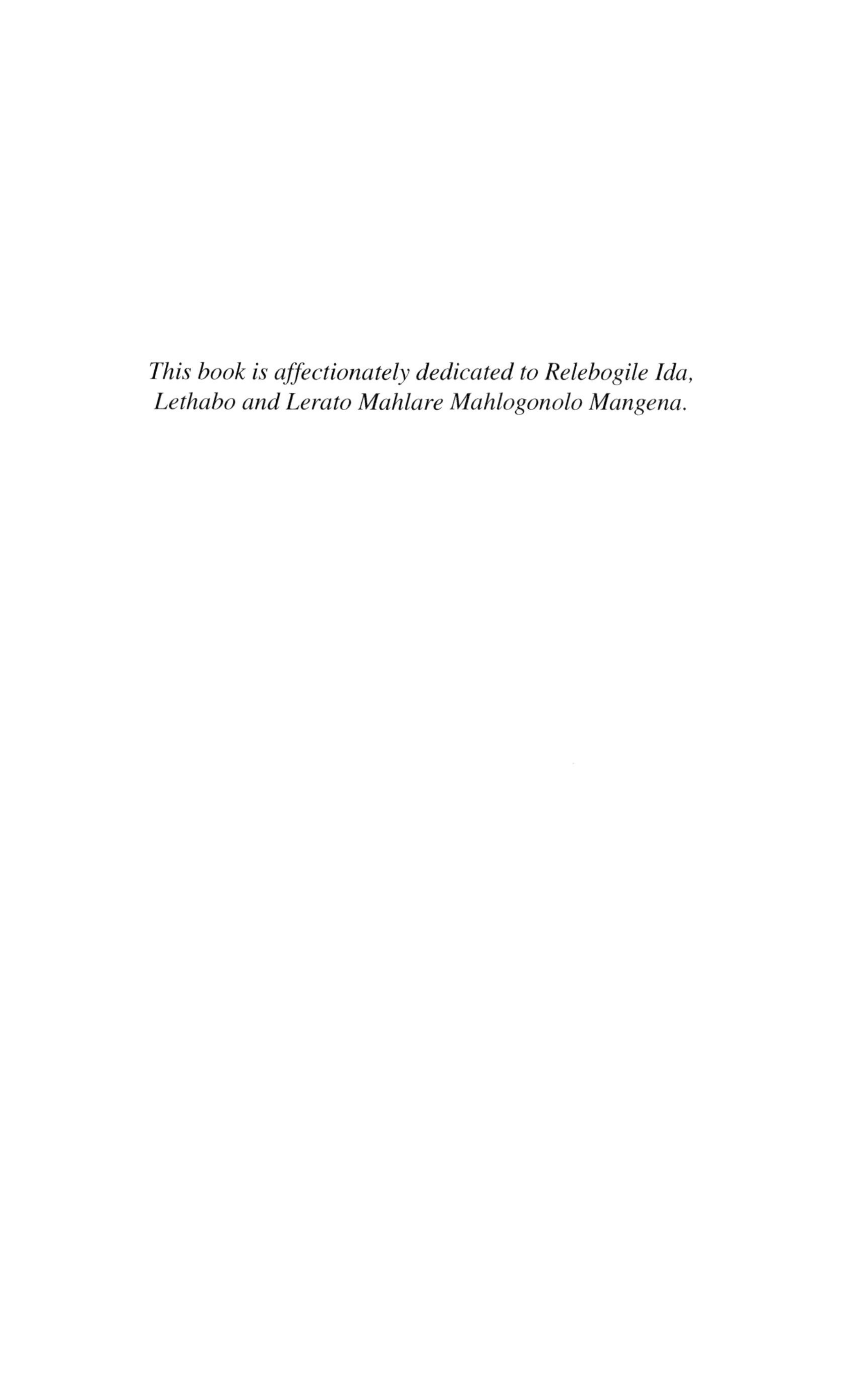

This book is affectionately dedicated to Relebogile Ida, Lethabo and Lerato Mahlare Mahlogonolo Mangena.

Contents

Preface

Glycine max (L.) Merrill, globally known as soybean or referred to by other states as 'soyabean', belongs to the order Rosales that is largely made up of many horticultural and ornamental plants. This Fabaceae species bears valuable dehiscent fruits botanically called legumes (pods), and each pod usually contains two to three seeds. Functional nutrients contained in these seeds make soybean one of the most important pulse crops worldwide. The crop owes its prominence to the grain it produces, whereby over 50 seeds bearing fruit pods can be produced per plant. This makes it one of the highest yielding pulses in agriculture. Soybean, like many other legumes, displays the most simplified architecture in its internal and external structures that are responsible for its many useful ecological and industrial services, including the crop's growth and development processes.

The plant's determinate, semi-determinate and indeterminate growth forms determine the morphological habit predominantly occurring in both early and late maturing groups of varieties that are widely cultivated for both subsistence and commercial farming. The shoot and root traits of this plant significantly differ in size or density according to the genotype and growth habit, and may also vary due to their response to environmental cues. Soybean accounts for over 146.23 billion US dollars of the total economic impact globally according to the 2021 Global Market Report by the International Institute for Sustainable Development (https://www.iisd.org.../ssi-global-market-report-soybean) and the Food and Agriculture Organisation of the United Nations (https://faostat.fao.org). This crop is among the few complete protein vegetable-based foods with almost all essential amino acids. As an oilseed crop rich in proteins (36%), carbohydrates (30%), oil (19%), crude fibre (5%) and ash (5%), it is mostly crushed into soy meal and oil used for the manufacturing of hundreds of edible and non-edible products. These soy-based products include animal feed, cooking oil, beverages, foods (cheese, milk, flour, etc.), pharmaceutical and other industrial products.

Amongst a list of obstacles to cumulative growth and crop yield, soybean is prone to pest attacks and remains highly sensitive to drought stress. Climate change-related problems such as increasing temperatures and more erratic rainfall exacerbate the spreading of diseases, often compromising crop productivity. There is a great need to ensure food and biofuel security whilst minimising the costs of production, environmental degradation and emission of greenhouse gases. Recombinant DNA technology and genetic engineering of crops using plant transformation techniques made it possible to isolate, characterise and manipulate genes to create new genetic combinations that are of value to agriculture, medicine and industry. Since the focus of genetics is the manipulation of genes, the fundamental goal of laboratory geneticists is to use these altered genes to contribute to agriculture in two important ways: (1) the production of nutritionally enhanced crops, and (2) the development of stress-resistance plant varieties. In soybean, recombinant DNA technology via applications such as *Agrobacterium tumefaciens*-mediated genetic transformation has been made possible by the construction of plasmid vectors used in the production of genetically modified plants. These plants contain genes that confer desirable growth characteristics.

The most prominent attributes include disease resistance, herbicide tolerance and a variety of mechanisms that are used by plants to favourably respond to abiotic stresses. This book, therefore, compiles critical information involving soybean transformation starting from the overview of genetic transformation in plants to photographic demonstrations of *in vitro* culture stages and processes performed by researchers during *Agrobacterium*-mediated genetic transformation. The first chapter provides a detailed overview of the purpose, process and key critical factors influencing the rate of transformation in soybean. Chapters 2 and 3 provide the reader with historical background accounting for the discovery of *A. tumefaciens* and its mediated DNA transfer mechanisms, with evidence showing that indeed this bacterium is behind the cause of crown gall disease in plants. Information and protocols/ techniques presented in Chapters 4 and 5, with extension to Chapters 6 and 7, are intended to comprehensively introduce *in vitro* plant tissue culture-based soybean transformation and other key alternative techniques to students and newcomers enthusiastic about plant genetic engineering via *A. tumefaciens* or genetic manipulation of plants using modern biotechnological tools in general.

Other chapters are dedicated to the development of stress-resistant plants and current and future applications of biotechnology in agriculture. *Agrobacterium*-mediated gene transfer still represents a huge milestone accomplished in the era of genetic engineering and modern plant molecular breeding. This tool serves as one of the most affordable, efficient and practical approaches utilised in crop improvement programmes. However, new tools that have emerged such as CRISPR/Cas gene editing appear to be more effective and promising. But, looking at the pace at which this technology is advancing, high prohibitory costs and slow public acceptance of agricultural innovations, new tools like genome editing may also experience a slow pace of adoption or suffer public scrutiny like in plant genetic transformation.

Phetole, Mangena
Department of Biodiversity
University of Limpopo
Republic of South Africa

Acknowledgements

Thanks to the many colleagues and students who gave support, words of encouragement and served as reviewers to some of the topics covered in this book. Thanks to Prof T. G. Mandiwana-Neudani, Prof S. M. Mahlo, Dr H. J. DuPlessis, Dr P. W. Mokwala, Ms Mabulwana T. P., Ms Mokgadi Modiba, Mr F. Nukeri, Mr Peter Mokumo, Mr B. Mdaka, Ms Tshepiso Ramalepe, Samuel Peta and Pirtunia Mushadu for providing support in a number of ways. My deepest appreciation also goes to the National Research Foundation of the Republic of South Africa, and the Department of Research Administration and Development of the University of Limpopo (Dr Mabila, Sis Noko, Mrs Hattingh, Mr Lekalakala, Mrs Kellermann, Moleke Heritage and John Mamokhere). Thanks to the Department of Biodiversity, School of Molecular and Life Sciences, Faculty of Science and Agriculture.

Funding for the ongoing research in the genetic transformation of soybean using *Agrobacterium tumefaciens* was provided by the National Research Foundation (Grant UID: 129747).

Author

Phetole Mangena is a senior lecturer and researcher of plant biotechnology at the Department of Biodiversity, University of Limpopo, Republic of South Africa. He has extensive experience in basic and applied plant sciences, specialising in plant genetic improvement of recalcitrant legumes for resistance to biotic and abiotic stresses. Mangena graduated with a doctoral degree (PhD) in Botany (Plant Biotechnology). He has authored and co-authored numerous scientific papers and book chapters relating to genetic improvement, morpho-physiological responses of plants to environmental stress and plant tissue culture. He is a sole editor of the book volume series 'Advances in Legume Research: Physiological Responses and Genetic Improvement for Stress Resistance' (Bentham Books) and the book *Legumes: Nutritional Value, Health Benefits and Management* (NOVA Science Publishers). Dr Mangena is a fellow-member of the World Researchers Association, editor for the *Research Journal of Biotechnology*, assistant editor of the *Journal of Environmental Biology and Agricultural Science* and editorial board member of the *Plant Biotechnology Persa*. Mangena is also a member of the South African Association of Botanists and the International Society for Horticultural Science. Dr Mangena received numerous honours and awards, including 1st Prize in the Innovation Hub GAP Bioscience Competition (Pretoria, South Africa) and the Research Excellence Award for Next Generation Researchers bestowed by the National Research Foundation of South Africa.

1 An Overview of Genetic Transformation in Plants

1.1 INTRODUCTION

Plant genetic transformation is defined as the alteration of the cell's genetic material by directly or indirectly incorporating exogenous DNA molecules from different sources across the cell membrane. In soybean, the first successful genetic transformation system was reported by Hinchee et al. (1988) using *Agrobacterium tumefaciens* strain pTiT37-SE harbouring a binary vector pMON9749. The plasmid vector contained kanamycin and glyphosate-resistant genes that were used as selectable makers to discriminate non-transgenic cells from host tissues that are transformation competent. All studies reported after the publication of Hinchee et al.'s experimental outcomes relied upon shoot organogenic cultures. In particular, cotyledons were used as explants to transfer and express antibiotic resistance genes for efficient transformation. However, the genetic transformation of soybeans still continues, particularly to develop novel cultivars that contain improved traits which guarantee better plant growth, flowering and yield, estimated through the crop's production values for the given harvest period as indicated in Table 1.1.

Unlike traditional breeding, plant transformation is a system valued because it exerts a high degree of culture control over each aspect of crop improvement, especially through the control of gene copy number and environmental conditions that accelerate genetic manipulations in many commercially viable crops, as well as recalcitrant plant species. This procedure holds considerable potential value for major biotechnological applications that are highly required to counteract seed recalcitrance and stress effects against many food legumes grown globally like soybean (Table 1.1).

A key critical factor encouraging the cultivation of soybean in countries such as the United States of America, Brazil and China indicated in Table 1.1 is the use of genetically modified seeds that ensures increased yield under conditions exposed to both biotic and abiotic stress. However, traditional breeding methods have in the past helped to successfully develop crop varieties with considerable improvements. The need for plant transformation has become greater due to its feasibility, rapidity and flexibility in terms of novel DNA isolations, introgression and laboratory operations required for the subsequent recovery of genetically modified plants.

The abovementioned advantages form part of a set of benefits that are not realised when using conventional breeding methods. Other disadvantages present in the conventional breeding but circumvented by plant transformation include high heterozygosity, auto-incompatibility of genes, limited genetic stocks and the simultaneous

DOI: 10.1201/b22829-1

TABLE 1.1
World average yields, area and production of leading grain crops for 2018–2020 projections

Crop	Country	Yield	Area	Production	GMO
Cotton	Brazil	12.8	1.6	0.00175	++
	China	27.8	3.5	0.00175	+
Corn	Brazil	100.0	17.5	5.18	++
	Russia	77.0	57.8	2.64	–
	Algeria	4.0	19.2	1.74	–
	Tunisia	1.5	0.6	1.72	+
	Morocco	2.9	0.1	2.19	+
	Syria	4.8	1.6	1.45	+
	Iraq	4.8	2.4	2.72	+
	Iran	16.8	6.7	1.81	–
	India	100.0	29.8	2.28	+
Rice	China	7.03	30.2	148.5	+
	India	3.91	44.5	116.0	+
	Indonesia	4.79	12.2	37.1	–
	Brazil	6.29	1.75	7.48	++
	Egypt	8.78	0.46	2.8	+
	United States	1.18	8.62	7.12	++
Soybean	United States	123.7	35.7	3147.9	++
	Brazil	3.24	36.1	117.0	++
	China	1.89	8.4	15.9	+
	India	1.05	11.0	11.5	+
	Canada	1.47	2.74	4.03	+
	Russia	1.47	2.74	4.03	–
	Indonesia	1.27	0.41	0.52	–
	Serbia	2.84	0.22	0.63	–
	Mexico	1.25	0.19	0.34	+
	South Africa	1.75	0.73	1.28	+
	Iran	2.29	0.07	0.16	–

Note: Yield is quantified in million metric tons, area in million hectares and production measured in million metric tons per hectare. Countries that allow the cultivation of genetically modified crops (+), expanded the area cultivated with genetically modified crops (++), restrictions and ban of import, distribution and commercialisation of genetically modified crops (–).

Data sourced from the United States Department of Agriculture, World Agricultural Production 5-19 (LOC 2020).

expression of multiple alleles (desirable/non-desirable alleles or in combination). Consequently, the generation of a breeding population that is highly variable for traits that are agriculturally interesting generally takes a long time, and there are no guarantees in the segregation of alleles (traits) into new and potentially better combinations of offspring. However, traditional plant breeding methods have been very successful in providing the required amounts of foods that have nourished the world population for many decades (Manshardt 2004; Datta 2007). The rapidly changing climate and its associated catastrophic events, including severe changes in weather patterns that lead to the disappearance of freshwater ecosystems, only accentuate the need for plant transformation for use in the development of varieties showing tolerance to abiotic stresses such as drought.

Apart from plant trait improvement, one of the main goals of plant transformation is to feed the world population and provide other beneficial industrial services. But there is scepticism and disagreements among consumers and researchers over the use of plant transformation to accomplish the genetic improvement of crops. Critics continue to ignore the novel traits, such as built-in resistance to several biotic and abiotic stresses, improved nutritional qualities and the provision of industrially useful secondary/primary metabolites. Furthermore, these advances provide genetic resources for plant breeders to access a new and broader gene pool (Datta 2007), and must be considered as one of the strides to genetically improve agronomically important legume crops.

1.2 MOLECULAR MECHANISM FOR DNA INTROGRESSION

As indicated in the introduction above, genetic transformation involves the uptake of exogenous DNA from the surroundings and expressing it in the cell's genome of the host to acquire a newly altered heritable phenotype. According to Smith et al. (1981), bacteria are considered to be the most naturally transformative cells in nature, with the exception of yeast and other eukaryotic organisms. Transformation is currently synonymous with the uptake of DNA by bacterial and plant cells, often with all cells involved being made competent either by chemical or physical means. Swords (2003) defined competence as the ability to accept linear or circular exogenous DNA dependent on the specific uptake system. For bacterial species showing natural competence, this phenomenon refers to the ability to take up and propagate plasmid DNA often without sequence specificity for the uptake system.

In the chemical transformation of bacteria in particular, divalent cations such as calcium chloride ($CaCl_2$) are used to facilitate DNA uptake across cellular envelopes and internalise it in the host cells as a stable genetic material. Alternative methods, which include the use of dimethyl sulfoxide (DMSO) and polyethylene glycol (PEG), other than electroporation used as a physical method have been successfully tested in different bacterial and plant species to induce cell competency (Asif et al. 2017). Electroporation uses a short electric pulse to disrupt cell structural integrity, making the membrane permeable to foreign DNA materials. This technique, together with particle bombardment, are typically used for the transformation of plant cells. Gold or tungsten particles coated with the DNA of interest are physically forced into the cells using a propeller referred to as a gene gun.

However, the past two decades were dominated by the discovery of both *Agrobacterium tumefaciens* and *Agrobacterium rhizogenes* Gram-negative soil-borne bacteria to transfer DNA to plant cells for the purpose of genetic engineering. Since the initial reports in the early 1980s preferably using *A. tumefaciens* to generate transgenic plants, scientists have attempted to establish a routine protocol for crop, medicinal and ornamental plant transformation. Gelvin (2003) reported *Agrobacterium*-mediated plant transformation as a highly complex process due to the evolving transformation processes involving specific strains and plant genotypes. Comprehensive insights into the fundamental biological principles guiding the routine transfer, expression and heritability of the genes, both for the host plant cell and the pathogen, are key to extending the utility of *Agrobacterium*-mediated transformation for biotechnological purposes. Currently, this approach is the most preferred and a widely used direct gene transfer technique across the globe.

The ability of *Agrobacterium* to encode transport proteins that play an essential role in the transmission of transfer-DNA (T-DNA) to plant cells using the virulence genes served as a breakthrough on how nucleic acids are translocated across biological membranes. Both the conductance channel and conjugal pores for the transmission of DNA substrates across the donor cell envelope served a vital role in enabling *Agrobacterium* use for genetic engineering in plants. The mechanism entails a bacterial transport system responsible for the delivery of oncogenic DNA (T-complex) across the bacterial envelope and to recipient host cells using a contact-dependent process. The transfer and expression of oncogenes from bacterium to susceptible plant cells characterise a neoplastic transformation taking place at the site of infection. Infection is initiated by an array of signals that include phenolic compounds, monosaccharides, acidic pH and low phosphate level present in a plant's wound site.

This process, signal perception, is regulated by the virulence genes (*VirA–VirG*) transduction system together with a periplasmic sugar-binding protein (Christie 1997). Although a significant contribution has been made in shedding light on molecular mechanisms of transformation, numerous laboratories are still working on ensuring that *Agrobacterium*-mediated transformation is routinely applied across genotypes. Issues such as genotype specificity, heavy reliance on certain bacterial strains and protocol inefficiencies, including a lack of standardised procedures, are still some of the major challenges plaguing the system. Furthermore, more insights could be gained by testing and refining the pathogen-host plant cell relationship and determining the most suitable regeneration conditions for the successful establishment of transgenic plants.

1.3 TISSUE TARGETING FOR PLANT TRANSFORMATION

Although plant transformation remains a method of choice in biotechnology for the improvement of most economically important plants, many species, including elite varieties, show immense recalcitrance to this technology. The exploitation of the Ti-plasmid system for plant cell transformation does not only rely on the successful development of competent bacterial cells containing a DNA construct, promoters that drive gene expression and introduction of the foreign DNA into plant chromosomes, but also involves a wide range of factors. Among these, the targeting of certain plant tissue as positive recipients of the T-DNA may also increase or influence the efficacy of the process. Gelvin (2003) suggested that the wide host range nature of *Agrobacterium* to transform gymnosperm, monocotyledonous and dicotyledonous plants, animals as well as fungi may not be a problem. But the report implied that T-DNA transfer to the recipients may not be a delinquent given the wider host range of the bacterium.

However, the attainment of highly efficient genetic transformation of cotton, some cereals, tree species and other many legume species of horticultural and industrial importance that remain a challenge proves otherwise. Thus, among the most challenging constraints, the targeted tissue explants are one of the most important factors. The ability of plant tissue that took up transforming DNA to heritably multiply and form various organs *de novo* remains a major goal and practical utility for plant transformers. This *de novo* formation of plant organs *in vitro* or *in vivo* using meristematic or non-meristematic tissues is termed organogenesis. Additionally, organogenesis has long provided the basis of asexual plant propagation largely from non-meristematic somatic cells.

The coupling of organogenesis with DNA transfer and expression for *in vitro* or *in vivo* multiplication of successfully infected and DNA incorporated pre-existing shoot meristems followed by *de novo* root formation of resultant microshoots or possible shoot regeneration on *in vitro* cultured tissues provided a premise that transgenic plants could be easily attained. In its broadest sense, this process was defined by Trigiano and Gray (2005) as a development process that results in functional and mature organisms through direct or indirect developmental sequence as discussed later in this chapter.

1.4 SCREENING AND SELECTION OF TRANSFORMANTS

Plant transformation, like other cloning methods, requires protocol steps verifying the successful transfer and expression of the gene of interest. This tool must contain within the process steps that allow the screening of transformants for the positive identification of transgenic tissues/plantlets before continuation with other experimental methods. However, one of the difficult aspects of this technology is still the inefficient and often cumbersome screening of a large number of lines to identify desired transformants. Hinchee and colleagues produced transgenic plants using an *Agrobacterium*-mediated gene transfer system, which relied upon the regeneration protocol in which shoot organogenesis was induced on cotyledons of soybean genotypes. Plantlets were tested for gene insertions by culturing on shoot induction medium containing kanamycin (Hinchee et al. 1988). According to Nap et al. (1992), the use of antibiotic kanamycin resistance gene as a selectable marker is widely used and acceptable, accompanied by a full legislative clearance of this transgenic trait.

This study highlighted factors such as the high substrate specificity of the encoded enzyme, less harm posed by antibiotics to human/animal health and the environment and the physicochemical characteristics of the antibiotics that exclude the requirement of selective conditions in the environment. A large number of agriculturally important crops that showed amenability to genetic manipulation by means of *A. tumefaciens*-based recombinant DNA technology were mostly selected using kanamycin. Freitas-Astua et al. (2003) reported infiltration of tobacco leaves with 100 mg/ml of kanamycin to clearly differentiate between transgenic and non-transgenic plant leaves (Figure 1.1). However, the leaves of non-transgenic tobacco plants exhibited conspicuous chlorotic spots, while no spots were presented on transgenic plants, indicated in Figure 1.1. This reaction and outcomes were reported to be simple, fast, affordable and non-destructive, regardless of the age of plants tested and their results comparable with PCR-based detection systems.

An optimised screening process that efficiently selects transgenic plants using integrated segregational analyses and molecular genetic tests (flow cytometry, diagnostic PCR and Southern blotting) was also reported. This workflow is illustrated on Figure 1.2, with quantitative PCR used in detecting changes in transcript accumulation confirming the overexpression or silencing of genes in the selection process (Gase et al. 2011; Cuhra 2015). However, following this kind of screening processes (shown in Figure 1.2) will be very costly and time consuming to say the least. For instance, purchasing a flow cytometric system to conduct this procedure will cost an average of 600,000 to 2 million in South African rands (~3,855,401 to 1.2 million USD). Screening remains one of the most critical challenges in creating and selecting transgenic lines that fulfil research objectives. Based on the available literature and protocols, kanamycin resistance genes are the most widely available and affordable

FIGURE 1.1 Tobacco leaves infiltrated with 100 mg/ml kanamycin as reported by Freitas-Astua et al. (2003) between transgenic (A) and non-transgenic (B) leaves.

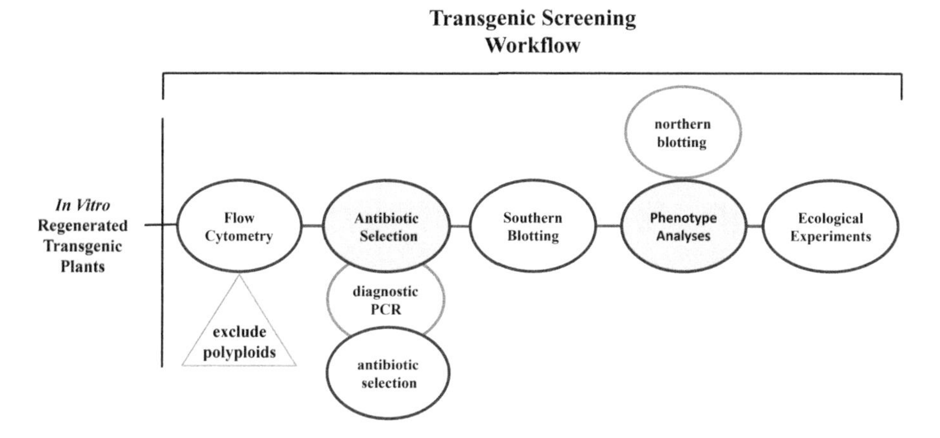

FIGURE 1.2 A workflow of efficient screening of transgenic plants for ecological research. The workflow can be repeated for the selection of positively identified plants, rapidly and reliably from T_0 to T_3.

efficient selection system applied in the transformation of many plant species carrying valuable agronomic interest.

1.5 TRANSFER TO NATURAL ENVIRONMENT

The ultimate success of a plant transformation culture also depends on the ability to transfer and re-establish vigorously growing plantlets from a regeneration culture. Four basic stages have to be followed to propagate transgenic plants *in vitro* before a transfer to the natural environment (Figure 1.3). Those stages include the preconditioning of the explant source and preparation of explants, establishment of aseptic cultures, infection and co-cultivation of targeted plant tissue with *Agrobacterium* and the proliferation of shoots from the cultured tissue explants (Figure 1.3) (Mangena et al. 2017). The conditions required for each culture depend on the species and culture conditions suitable for the micropropagation of genetically altered plants. This method provides genetic stability and ease of culture attainment for many plant species amenable to the chosen transformation protocol. Consequently, the culture establishment method adhered to before acclimatisation to *ex-vitro* conditions may play an important role in the development and use of this approach to produce millions of clones or genetically manipulated plants annually (Trigiano and Gray 2005).

But among the many challenges facing *Agrobacterium* gene transfer technology, regenerated plantlets remain difficult to acclimatise despite all hardening procedures being followed. Poor survival rates frequently encountered are due to the fact that regenerated plantlets are still under a heterotrophic mode of nutrition and possess very poor control of water loss as a result of an under-developed dermal tissue system. Essentially, *in vitro* grown plants are very tender and difficult to grow under *ex vitro* conditions due to the high humidity in the culture vessels and other growth room conditions (controlled temperature, light intensity and photoperiod) (Hazarika 2006). Every regeneration and transformation protocol should be accompanied by

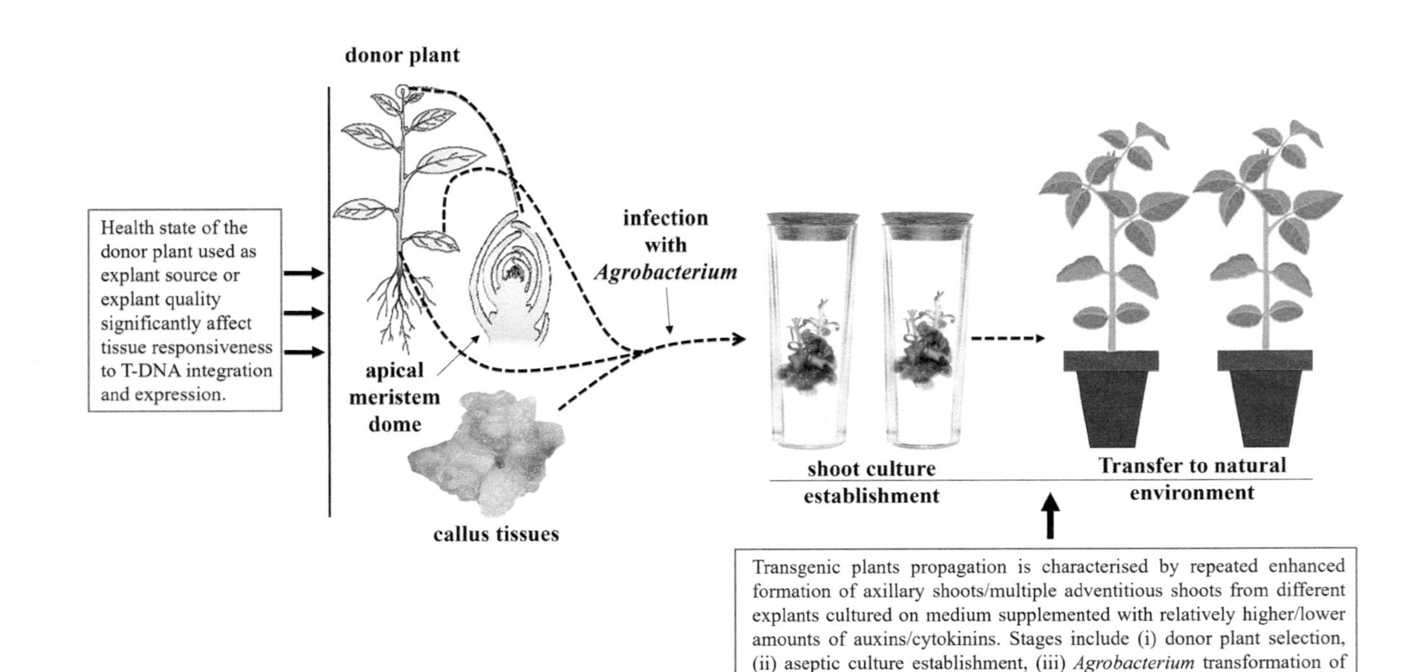

FIGURE 1.3 Typical illustration of the *in vitro* regeneration of plantlets for subsequent transfer to natural environmental conditions. This micropropagation system can be used for both transgenic and non-transgenic plant establishment to overcome limitations faced during hardening or acclimatisation.

efficient stages that facilitate the efficient transplantation to field environments. Relief measures that prevent the high percentages of lost plants and damage or severe abnormalities could improve widespread use of *in vitro* regeneration even for cloning when plants are successfully transferred to *ex vitro* conditions.

1.6 SUMMARY

Plant genetic transformation of crops represents a major milestone in modern agriculture, especially for crop breeding against biotic and abiotic stress. Advances made thus far have raised expectations that highly nutritious foods and more stress-resistant crop varieties will be developed using this technique. But, due to the nature of culture condition challenges, recalcitrance and lack of reproducibility and genotype-independence in some protocols, the technology has not yet fully translated into the development of genetically improved cultivars (Mangena 2020).

As the improvement of soybean germplasm through conventional breeding remains severely challenged, tools like genetic engineering must be at the forefront of developing means for achieving high yield, increased proteins, oils and vitamins for crops like soybean, particularly, in combating food insecurity in developing African countries. It should be noted, however, that in most instances these newly developed varieties are preferably used to improve plant growth and yield under unfavourable cultivating conditions, particularly because soybeans have proved to be highly susceptible to drought and pests, with abiotic stress serving as the greatest limiting factor. Finally, advances in plant transformation methods are still highly required to continue helping in crop improvement, the production of useful secondary metabolites, herbicide resistance and environmental stress tolerance and for the creation of the much anticipated genetic variants from existing orphan legume crops.

In this technique, the amenability of tissue explants, reproducibility of the protocol and recovery of competent cells with T-DNA expressions are the mainstay in ongoing endeavours to develop the most effective and efficient protocols for genetic improvement in soybeans and other legume species (Kado 2014). Therefore, the research trajectory in plant tissue culture-based genetic improvement technology must tackle and strengthen the application of plant transformation to advanced mechanisms of transgenesis in the development of the much-needed new crop cultivars as indicated above.

1.7 ABBREVIATIONS

$CaCl_2$ Calcium chlorite
DNA Deoxyribonucleic acid
DMSO Dimethyl sulfoxide
GMO Genetically modified organisms
LOC Library of Congress
PCR Polymerase chain reaction
PEG Polyethylene glycol
T-DNA Transferred DNA
USD United States dollars
Vir Virulence

REFERENCES

Asif A, Mohsin H, Tanvir R and Rehman Y. (2017). Revisiting the mechanisms involved in calcium chloride induced bacterial transformation. *Frontiers in Microbiology* 8(2169), 1–5.

Christie PJ. (1997). *Agrobacterium tumefaciens* T-complex transport apparatus: A paradigm for a new family of multifunctional transporters in Eubacteria. *Journal of Bacteriology* 179(10), 3085–3094.

Cuhra M. (2015). Review of GMO safety assessment studies: Glyphosate residue in Roundup Ready crops is an ignored issue. *Environmental Science Europe* 27(20), 1–14.

Datta SK. (2007). Impact of plant biotechnology in agriculture. In Nagata T, Lorz H and Widholm JM (eds), *Biotechnology in Agriculture and Forestry*. Springer-Verlag, Berlin Heidelberg. pp. 1–6.

Freitas-Astua J, Astua-Monge G, Polston JE and Hiebe E. (2003). A simple and reliable method for the screening of transgenic tobacco plants. *Pesquisa Agropecuaria Brasíleira* 38(7), 893–896.

Gase K, Weinhold A, Bozorov T, Schuck S and Baldwin IT. (2011). Efficient screening of transgenic plant lines for ecological research. *Molecular Ecology Resources* 11(5), 890–902.

Gelvin SB. (2003). *Agrobacterium*-mediated plant transformation: The biology behind the "gene jockeying" tool. *Microbiology and Molecular Biology Review* 67(1), 16–37.

Hazarika BN. (2006). Morpho-physiological disorders in *in vitro* culture of plants. *Scientia Horticulturae* 108, 105–120.

Hinchee MAW, Connor-Ward DV, Newell CA, McDonnell RE, Sato SJ, Gasser CS, Fischhoff DA, Re DB, Fraky RT and Horsch RB. (1988). Production of transgenic soybean plants using *Agrobacterium*-mediated DNA transfer. *Bio/Technology* 6, 915–922.

Kado CI (2014). Historical account on gaining insights on the mechanism of crown gall tumorigenesis induced by *Agrobacterium tumefaciens*. *Frontiers in Microbiology* 5, 340.

Library of Congress (LOC). (2020). Restrictions on genetically modified organisms research and reports. http://www.loc.gov. Date accessed: 23/09/2020.

Mangena P. (2020). Genetic transformation to confer drought stress tolerance in soybean (*Glycine max* L.). In Guleria P, Kumar V and Lichtfouse E (eds), *Sustainable Agriculture Reviews 45- Legume Agriculture and Biotechnology*. Springer Nature, Cham. pp. 193–224.

Mangena P, Mokwala PW and Nikolova RV. (2017). Challenges of *in vitro* and *in vivo Agrobacterium*-mediated genetic transformation in soybean. In Kasai M (ed), *Soybean-The Basis of Yield, Biomass and Productivity*. IntechOpen, Rijeka, Croatia. pp. 75–94.

Manshardt R. (2004). Crop improvement by conventional breeding or genetic engineering: How different are they? *Biotechnology* BIO-5, 1–5.

Nap JP, Bijvoet J and Stiekema WJ. (1992). Biosafety of kanamycin-resistant transgenic plants. *Transgenic Research* 1, 239–249.

Smith HO, Danner DB and Deich RA. (1981). Genetic transformation. *Annual Review of Biochemistry* 50, 41–68.

Swords WE. (2003). Chemical transformation of *E. coli*. In Casali N and Preston A (eds), *E. coli Plasmid Vectors: Methods in Molecular Biology*. Humana Press, Totowa, New Jersey. pp. 49–53.

Trigiano RN and Gray DJ. (2005). *Plant Development and Biotechnology*. CRC Press, Boca Raton. pp. 211–251.

2 Plant Transformation History

2.1 DISCOVERY OF *AGROBACTERIUM TUMEFACIENS*

Agrobacterium tumefaciens was first discovered by Smith in 1892. Erwin Frank Smith discovered this bacterium after noticing fleshy outgrowth developments on the roots of several fruit-bearing trees (Zeanen et al. 1974; Somssich 2019). Fruit trees on which these 'plant cancers' were predominantly noticed included species of *Malus* (apple), *Prunus* (apricot, cherry and plum) and *Vitis* (grapes). Other plants that also displayed such infectious outgrowths were *Populus* (cottonwood), *Salix* (willow) and *Rosa* (roses), which form part of a group of many plants that showed susceptibility to this bacteria-induced gall disease. But most of the infected and affected species were members of the rose family as listed above. The literature indicates that crown gall disease causes rough, woody tumour-like galls on the roots, trunks and occasionally branches of more than 600 species of angiosperms. Amongst these flowering, fruit-bearing plants are common vegetables and weeds. Further research showed that this disease impedes the proper functioning of the vascular tissue system by interrupting the flow of dissolved mineral nutrients and water, as well as the translocation of photoassimilates in the phloem.

According to Poncet et al. (1995) young plants heavily contaminated with this disease are severely affected and can be subsequently killed. Evaluations of the impact of crown gall disease on plant development revealed severity depending on factors like the conditions of growth, degree of resistance/susceptibility, time of infection, type and status of infected organs and the rate of opportunistic infections. The mystery was, however, soon unravelled when Erwin F. Smith began his detailed work on the fleshy outgrowths of the crown galls in the late 19th century. Smith, a pioneer in plant pathology, first reported on what was later termed 'plant cancer' when he was inquisitive about the aetiology and control of crown galls, and their morphology. According to Campbell (1983), Smith's fascination with crown gall disease was fuelled by his conviction that animal cancers were comparable in certain respects to plant tumours, typically on the manner of causation of the diseases or tumour conditions.

The idea of tumour formation, potential stimuli and growth inhibition grew from Smith's ambitions of speculation on this bacterial disease. His work continued to strengthen the argument and provide insights on the types of plant tumours, wherein some were formed spontaneously without intervening pathogenic microorganisms and others induced by infectious bacteria, viruses and fungi, as well as insects, etc. Currently, it is well known that plant tumours, either caused by pathogenic bacteria

DOI: 10.1201/b22829-2

or spontaneous, in various tissues of plants both involve phytohormonal imbalances which influence the induction of abnormal growths of meristematic tissues.

Meristems are groups of undifferentiated parenchyma cells that continuously undergo rapid plant cell division. In 1977, Chilton and colleagues confirmed that *A. tumefaciens* was the bacterium inciting this formation of crown gall tumours caused by incorporating part of the virulence circular (plasmid) DNA carried alongside its binary vector. The study identified the presence of tumour DNA using a labelled plasmid DNA analysed through its sequencing and gel electrophoresis. According to the findings in this report, about 40% of the DNA sequence copies responsible for inducing tumours were available per diploid tumour cell, at a rate of about 3.7×10^6 Daltons (Da) of genetic material content (Chilton et al. 1977). These findings and other related studies served as the most critical developments that brought about clarity on whether *Agrobacterium* was a causative agent of this crown gall disease.

The idea that bacteria could infect plants was initially understood as the most outrageous hypothesis by many microbiologists at the time, particularly by Smith's big opponent Alfred Fisher. Alfred was a highly reputable expert in the field of microbiology with many bacteriology publications since around 1879. This evidence later concluded the then ongoing reprehensible debate about plants infecting bacteria amongst scientists at the time with Chilton's report providing empirical evidence answering the big question of whether bacteria could be the cause of a crown gall disease in plants. However, earlier in 1907 before these findings were made by Chilton et al. (1977), Erwin Smith had already published evidence showing that it is indeed a bacterium that causes these tumours and proposed the name *Agrobacterium tumefaciens* (Smith and Townsend 1907).

Apart from the report by Chilton et al. (1977), Shokraii and Azizian (1979) also surveyed and comparatively analysed tumourigenesis effects of *A. tumefaciens* on some plant species in Iran. Their findings also revealed that tumourigenesis was species-specific and confined to plant organs such as stems and roots, but was absent or not very common in mature leaves. Furthermore, the study clearly illustrated that phytohormones like indolyl-3-acetic acid (IAA) were more important for the initiation and enhancements of primary as well as secondary tumours induced by *Agrobacterium*. However, all developments became possible to achieve subsequent to Smith's full description and details of tumours caused by this bacterium published in 1912, which was about 20 years after making his initially denounced speculative observations.

2.2 TUMOUR-INDUCING PRINCIPLE

Generally, *Agrobacterium tumefaciens* uses its unique ability to transfer its genomic segments to transform plant cells and cause tumours. All the genes required for this process are carried by larger tumour-inducing plasmids (Ti) and binary DNA found within the bacterial cell cytoplasm. In a nut-shell, infected plant cells overproduce phytohormones that cause uncontrollable cell proliferation resulting in crown galls. A simplified overview of this transformation process leading to the uncontrolled cell proliferation on infected plant cells is illustrated in Figure 2.1. According to this

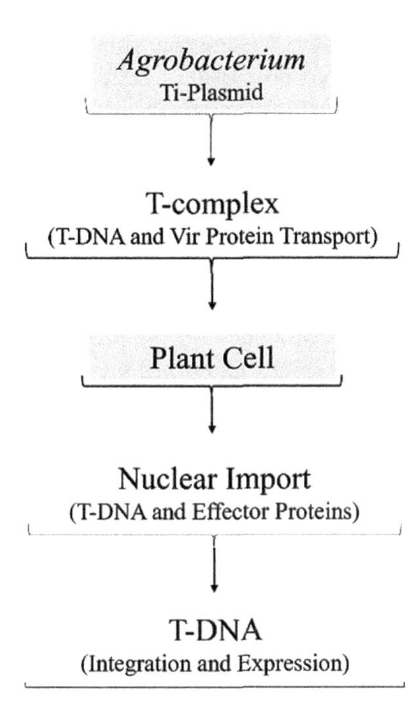

FIGURE 2.1 Brief summary of stages involving the intracellular movement of T-DNA for genome integration and expression. *A. tumefaciens* containing two genetic components located on the tumour-inducing (Ti) plasmid are required (T-DNA and *Virulence* (*Vir*) region), including bacterial attachment to the plant cell wound using chromosomal virulence (*chvA*, *chvB* and *pscA* genes). This is followed by cytoplasmic and nuclear import of T-DNA and effector proteins for integration and expression.

process, *Agrobacterium* transfers its segment of DNA (T-DNA) for incorporation in the host's genome after gaining entry and then infecting plants through wounded cells. This bacterium senses sugars and phenolic acids produced by wounded host tissues, and uses the bacterial type IV secretion system (T4SS) to transfer its genetic materials into the hosts (Hwang et al. 2017).

This neoplastic tumour-like cell growth is induced by the expression of oncogenes found specifically in the transforming T-DNA of the Ti-plasmid. This DNA is transported from bacterial cells into the plant's nucleus and becomes nucleus integrated wherein it targets host cells' chromatin material to achieve its staple integration. According to Gelvin (2012), this fundamental mechanism of pathogenesis remains the same throughout, particularly for all *Agrobacterium* species including both *A. tumefaciens* and *A. rhizogenes*, causative agents for hairy root disease, as well as *A. vitis* and *A. rubi* that cause galls in grape plants and sugarcane, respectively.

Earlier evidence suggested that it is the various proteins associated with the T-DNA that were responsible for controlling the trafficking of the genetic materials that are transferred into the host plant cells (Chilton et al. 1977). This then became more apparent when it was recently reported by Gelvin (2012), Hwang et al. (2017)

and many other researchers that border sequences and virulence proteins of the bacterium function to mediate T-DNA processing transfer and delivery into transformed plant tissues. For the purposes of clarity and further deliberations, these elements will be discussed later in the subsequent chapters, especially Chapter 3 and again highlighted briefly in Chapter 8. Nonetheless, from the transformation stages represented in Figure 2.1, T-DNA integration and expression precede the tumourigenic state of cells. The integration of T-DNA into the plant cell genome overproduces phytohormones auxins and cytokinins, also known as plant growth regulators (PGRs), which eventually cause the cancerous phenotype (Escudero et al. 1995).

Other metabolic compounds produced during this process include saccharides, amino acids, carboxylic acids, opines and other several alkaloids that are used as crucial sources of nutrients nourishing the infecting bacteria. It is, however, important to note that not all metabolites are used as nutrients, but some, like polygalacturonases, were hypothesised to facilitate bacterial attachments and systematic colonisation. The production of polygalacturonases commonly involved in the degradation of pectin substances, also known as pectin depolymerase, was observed by Jung et al. (2016) in grapevine plants infected by *A. vitis*. These metabolites are also implicated in the facilitation of *Agrobacterium* cell growth, gene transfer and virulence induction that take place after the bacterium has sensed phenolic compounds, sugars and other nutrients secreted by wounded tissues on the plant.

Agrobacterium then uses its cellular machinery to mediate crown gall formation to create a better environment for its own survival. On the other hand, metabolic shifts are also experienced by *Agrobacterium* species living in different habitats, and the nutritional compositions of each environment. Typically, these bacteria will be faced with more nutrient limitations and environmental stress when living in soil and the rhizosphere outside than in the nodules or galls. Xu et al. (2020) reported on the importance of these metabolites for tumourigenesis. Their findings indicated that significant metabolic shifts negatively influenced this bacterium's response to pathogenesis and tumour formation.

2.3 STRUCTURE AND MODIFICATIONS OF THE *AGROBACTERIUM* CIRCULAR DNA

Agrobacterium's remarkable transformation abilities have revolutionised horizontal gene transfer in the past decades. Advances in bacterial genetics and nucleic acid chemistry, which started almost a century ago, changed the way plant genetics was perceived in relation to crown gall disease. This discovery led to the development of a whole new natural genetic engineering system that is now dedicated to the genetic manipulations of horticultural and ornamental plants. As indicated in previous discussions above, *Agrobacterium* research was first focused on the formation of crown gall disease in fruit trees, as well as some gymnosperms. However, insights by Armin Braun in 1958 helped shine a spotlight on this bacterium and the tumour-inducing concept that provided further information on pathogenesis, tumourigenesis and its potential application for beneficial DNA modifications in plants.

The discovery of an unusually large plasmid DNA, its associated crown gall disease and the introgression of DNA segments in host plant genomes made the use of *A. tumefaciens* in the genetic engineering of crops seem more possible (Zaenen et al. 1974; Chilton 1977). The fact that this bacterium could cause the nutritional properties of infected plants to change paved the way for many scientists to raise intriguing plant transformation questions. From 1960, a number of laboratories started investigating how a piece of the plasmid DNA was transferred and randomly integrated into the host plant chromosomes and then analysed the protein metabolites encoded by such genes (Nester 2015). Ultimately, Kerr (1969) provided comprehensive insights demonstrating that indeed DNA transformation with *Agrobacterium* and *Rhizobium* was responsible for the virulence showed by these pathogens. This was done by deliberately inoculating tomato (*Lycopersicon esculentum*) plants with virulent strains of *Agrobacterium* which later showed the development of tumours, merely a few weeks after inoculation.

All reported evidence culminated into ample concrete evidence demonstrating that *A. tumefaciens* is capable of transferring its segments of plasmid DNA to cause genetic changes in host cells. It thus became apparent that replacing transferred tumour-inducing genes with some desirable foreign DNA fragments could allow for the introduction of any new exogenous DNA in plant cells. Scientists then started exploring both circular and linear *Agrobacterium* chromosomal DNA, as well as its virulence and metabolic parasitism, in order to manipulate the genomes of plants. Three major biovars (biovar 1, biovar 2 and biovar 3) of *A. tumefaciens* that were initially explored for the above purpose also differed according to the following:

1) Metabolic characteristics that include induced opines (octopine and nopaline) and lysopine;
2) Relationship of the biovars with other genera in the family Rhizobiaceae;
3) Host range specificity;
4) Genome architecture or chromosomal structure.

The differentiation of biovars was commonly based on the utilisation of different carbohydrates and some biochemical products. Reports showed that metabolites such as octopine are derived from the condensation product of arginine and pyruvic acid, nopaline is the product of arginine and alpha-ketoglutaric acid and lysopine is a condensation product of lysine and pyruvic acid (Nester 2015). These secondary metabolites were found to be produced as a result of the alterations of plant hormones following T-DNA integration and expression. Furthermore, this relationship was only established within *A. tumefaciens* and was not particularly relevant to the bacterium's pathogenicity except for biovar 3 which was found to be common in grapevines.

This biovar 3 strain has been spread across the globe through the planting of plant cutting stocks that were contaminated (Hwang et al. 2013). It ultimately became clear that the transfer, integration and expression of genetic materials from these bacteria are a characteristic biological mechanism of the Ti-plasmids rather than its chromosomal DNA. These revelations raised many challenging and interesting questions about the nature of oncogene expression encoded by the T-DNA.

2.4 GENETIC TRANSFORMATION IN PLANTS

The mechanism of plant transformation was not well understood until Garfinkel et al. (1981) constructed a genomic map of the transferred genetic materials, and provided further insights on the function of the virulence region (oncogenes) found on the co-resident Ti-plasmid. Transferred DNA regions were found to play a key role during plant transformation by functioning as the transfer sequences bordered by the left and right sequences of the T-DNA. In 1984, Michael Bevan became the first scientist to indicate that *Agrobacterium* DNA required some modifications before it could be used for nuclear genome transformation in plants. Bevan's (1984) report first indicated that the tumour-inducing genes have to be removed as their expression interferes with the regeneration of fertile transgenic plants. Secondly, selectable marker genes have to be inserted to prevent or decrease the proportion of cells containing chimeric genes that serve as pseudo-transgenic plants.

These developments led to the predominant use of kanamycin-resistant genes as the dominant selectable marker. Transgenic plants containing kanamycin resistance comprised of coding sequences derived from the bacterial neomycin phosphotransferase gene in addition to the gene of interest (*goi*). Currently, *Agrobacterium* strains utilised for plant transformation contain different vector constructs made up of such gene combinations. Any strain, such as 1D1108, 1D1487, 1D1478, 1D1460 and 1D132 isolated from *Euonymus*, apple, *Salix*, sugarcane and cherry, respectively, can be used for genetic transformation in plants (Hwang et al. 2013). Paz et al. (2006) reported derivative constructs of the base vector pTF101.1, a derivative of pPZP binary vector containing the right and left T-DNA border sequences from a nopaline strain with a broad host origin of replication (pVS1) and a spectinomycin-resistant marker gene (*aadA*) for bacterial selection.

The plant selection marker gene cassette of this construct consisted of (1) double 35S promoter (2 × P35S) of cauliflower mosaic virus (CaMV), (2) tobacco etch virus translational enhancer and (3) phosphinothricin acetyl transferase *bar* gene. The phosphinothricin transferase (PAT) proteins are encoded by the *bar* gene coding sequence derived from either *Streptomyces hygroscopicus* or *S. viridochromeogenes*. These genes are also present in glufosinate-ammonium herbicide-tolerant plant varieties of various crops such as corn, cotton, rice, rapeseed and soybean (Herouet et al. 2005). Earlier indications that bacterial T-DNA was actually transferred and integrated into the plant's genome resulted in molecular developments mentioned above. It was the findings of Zambryski et al. (1980) which determined that the highly conserved oncogenic plasmid sequences were flanked by the T-DNA on one side and flanked by plant DNA on the other side.

Boundaries of the T-DNA were cloned from tobacco plant cells transformed with *A. tumefaciens* which contained several direct repeats and also appeared to be tandemly repeated. According to this report, detailed mapping with restriction enzymes and nucleotide sequence analysis of two independent clones was used to study the molecular structure of T-DNA ends, subsequently leading to the full characterisation of the Left and Right Border sequences. At this stage, following further findings made by Zambryski et al. (1980) and Chilton et al. (1980), it was then clarified that the genetic transformation of plant cells could be achieved using the T-DNA and segments of *goi* without tumourigenesis. As previously highlighted, these efforts were mostly impeded by the malignant character of pathogenesis as it was very difficult to regenerate a healthy plant from transformed tissues as previously reported by Bevan (1984).

2.5 THE FIRST TRANSGENIC PLANT

After revelations that plant transformation systems required disarmed *A. tumefaciens* strains containing the full virulence genes to mediate T-DNA introgressions, the modification of Ti-plasmids resulted in the development of numerous plasmid variants (Chilton et al. 1980; Ooms et al. 1981; Hwang et al. 2013; Somssich 2019). This followed several publications which indicated that armed strains inhibited the *de novo* organogenesis of shoots and roots; meanwhile the newly developed plasmid variants induced small tumours and shoots during plant cell and tissue culture. The changes made on plasmid variants meant that the genes of interest also had to be incorporated within the alternative T-DNA region found in a binary vector. Otten et al. (1981) later reported the insertion of the bacterial transposon Tn7 in *Agrobacterium* T-DNA to obtain mutants of octopine Ti-plasmids without introducing genes conferring desirable traits.

This work then indicated that crown gall tumours induced on tobacco plants by an *Agrobacterium* strain carrying this particular mutant Ti-plasmid gave rise to vigorously growing adventitious shoots. Their evidence also showed that the tobacco shoots carrying the pGV2100 vector construct efficiently developed into morphologically and functionally stable plants whose offspring also showed gene transmission in a Mendelian fashion. According to Otten's report, at least 41% of these plantlets contained a transferred DNA-specific enzyme lysopine dehydrogenase (LpDH) that was used for the positive identification of transgenic tobacco plants, and these accomplishments were made with not enough tumourigenesis. In 1983, Herrera-Estrella and his colleagues introduced foreign genes into tobacco plant cells with a modified Ti-plasmid vector. This report illustrated the construction of an expression pLGV2381 vector construct from a nopaline synthase promotor and octopine synthase coding sequences, incorporated together with chloramphenicol acetyl transferase (*cat*) antibiotic-resistant gene (Herrare-Estrella et al. 1983).

This was deemed the first ever highly successful genetic transformation using the pLGV2381 Ti-plasmid of *A. tumefaciens* following the introduction of foreign genes into plant cells. Luis Herrera-Estrella documented some details of this major historical breakthrough in his recent biographical article titled 'My Journey into the Birth of Plant Transgenesis and Its Impact on Modern Plant Biology' where he also shared his personal experience in the development of the technology to produce transgenic plants in the Marc van Montagu and Jeff Schell Laboratory (Herrera-Estrella 2020). It was then revealed in Koncz et al. (1983)'s publication that this nopaline synthase promoter sequence is the one that efficiently enabled the expression of the *cat* transgene in tobacco. As the first transgenic plant was obtained, Bevan et al. (1983) also reported efficient transformation of *Nicotiana* variety Turkish using a Ti-plasmid carrying a Geneticin418 (G418) and selectively isolated the transformed cells using a Murashige and Skoog (MS) basal culture medium supplemented with 500 µg/ml carbenicillin.

Michael Bevan and his colleagues, including Chilton, used a neomycin (*neo*)-resistant gene derived from Tn5 coding for neomycin phosphotransferase II, and conferred resistance to the aminoglycoside antibiotic geneticin. These developments were then followed by reports made by researchers in the Monsanto Laboratory who published transgenic *Petunia* lines carrying the bacterial aminoglycoside-3'-phosphotransferase (*npt*) gene controlled by nitrous oxide (*nos*) regulatory sequences among others (Somssich 2019). Progress made did not slow down the pace of research into pathogenicity and tumourigenesis since a completely non-oncogenic Ti-plasmid was yet to be developed to transfer genes of interest into plant cells without causing any tumourous growths.

Nevertheless, evidence of such a plasmid was then published by De Block et al. (1984) who recombined nopaline synthase promotor and bacterial coding sequences with specific resistance to kanamycin, chloramphenicol or methotrexate, inserted in the first non-oncogenic Ti-plasmid vector pGV3850. Tobacco protoplast cells co-cultivated with this plasmid allowed for the selection of transgenic calli to regenerate phenotypically normal and fertile plants in tissue culture. Findings made in this study were then used by many researchers to model experiments involving *Agrobacterium*-mediated genetic transformation in many crop plant species. Among the many reports that followed after this further optimisation of the Ti-plasmids, Hinchee et al. (1988) reported the first genetic transformation in soybean using a non-oncogenic pTiT37-SE strain carrying the pMON9749 vector with kanamycin resistance and glyphosate tolerance. This aminoglycoside antibiotic resistance and herbicide tolerance were conferred by the *β-glucuronidase* gene well known as the *GUS* activity.

The study reported an overall 6% transformation frequency using a cotyledonary explant system on MS medium supplemented with 100 mg/l kanamycin in two soybean cultivars, Peking and Delmar. The demonstration of successful genetic transformation in soybean came soon after several genetic manipulation protocols were developed in many crop species such as tomato, rapeseed, cotton and flax as reported by McCormick et al. (1986), Fry et al. (1987), Umbeck et al. (1987) and Basiran et al. (1987), respectively. Even though the first successful genetic engineering of soybean by Hinchee et al. (1988) dates back to over three decades ago, improvements in the process are still necessary to achieve higher efficiencies. Thus, researchers continue to work immensely to develop organogenesis-based transformation methods using different *A. tumefaciens* strains and binary vector constructs containing different sets of genes, including genes for nutritional enhancements or to confer tolerance to biotic and abiotic stress factors. A transformation protocol that is capable of moving foreign DNA from a bacterium cell into tissues of soybean explants, thereby efficiently and rapidly altering its genome in a stable manner, remains highly required.

This task is a central goal in many research laboratories, particularly with the greater goal to enhance crop productivity and increase the diversity of the existing food base that the human population relies upon.

2.6 ROLE OF PLANT TISSUE CULTURE

However, genetic transformation via *Agrobacterium tumefaciens* is no longer an approach that is in its infancy. Unlike in most monocotyledonous plants, the transformation process in some dicotyledonous plants remains highly recalcitrant. Legume crops such as soybean still face many hurdles, and the frequency of transgene expression remains very low. Most researchers have, therefore, sought to improve and optimise this technique through the use of plant tissue culture. In order to overcome some of the challenges, researchers normally grow plant cells or tissues (callus and protoplasts) and organs (shoots, stems, embryos, flowers, seeds, etc.) under aseptic culture conditions on a sterile nutrient culture medium enriched with various organic supplements (amino acids, organic additives or PGRs/exogenous plant hormones) (Mangena 2021).

The system remains preferable because it exerts a high degree of control over each aspect of tissue culture by controlling culture media composition and environmental conditions to accelerate and successfully achieve *in vitro* plant regeneration. Plant tissue culture has been used alone by many commercial laboratories across the globe, without genetic transformation, for the clonal micropropagation of many commercially viable and recalcitrant plant species. But when plant tissue culture is coupled with molecular genetics, it then serves as a core technology for the genetic engineering of plants (Trigiano and Gray 2005). The first successful plant cell and tissue culture protocol was reported by Gottlieb Haberlandt, an Australian botanist, about 120 years ago, at the turn of the 20th century. Gottlieb, the son of a European 'soybean' pioneer who introduced soybean cultivation in Western and Central Europe, first attempted the use of cell cultures in plants, ushering in a whole new era of plant regeneration by stimulating later investigations (Krikorian and Berguam 2003).

This was a remarkable achievement at the time, leading to the many more accomplishments reported to date on the optimisation and application of plant tissue culture protocols. Currently, plant development and totipotency are achievable using PGRs (listed in Table 2.1) needed for cell division, cell proliferation and embryo induction on a modified culture medium. PGRs can cause dramatic effects in a culture at lower to intermediary concentrations (0.001–10 μM) and ratio (intermediate auxin–cytokinin ratio, low auxin–cytokinin ratio and high auxin–cytokinin ratio). Based on these compositions, the already mentioned ratio of PGRs can be used to promote callus, shoot and root formation, respectively, in a culture. The discovery of the first PGR indole-3-acetic acid (IAA) (whose molecular weight is indicated in Table 2.1) in 1937 by Kenneth Vivian Thimann opened new avenues at the time for rapid progress and to achieve the first successful culture of plant tissues (Trigiano and Gray 2005).

Kenneth was one of the world's leading botanists, and he died peacefully at his home on Wednesday, January 15, 1997, at the Quadrangle, Haverford, Pennsylvania.

TABLE 2.1
Plant growth regulators (PGRs) commonly used for plant tissue culture, their abbreviations and molecular weights (MW)

Plant growth regulators	Abbreviations	MW (g/mol)
Auxins:		
Indole-3-acetic acid	IAA	175.2
Indole-3-butyric acid	IBA	203.2
1-Naphthalene acetic acid	NAA	186.2
2,4-Dichlorophenoxyacetic acid	2,4-D	221.04
4-Amino-2,5,6-trichloropicdinic acid	Picloram	241.5
Cytokinins:		
6-Benzylamino purine/benzyl-adenine	BAP	225.2
Kinetin	Kin	215.2
Zeatin	Zea	219.2
Thidiazuron	TDZ	220.3
Abscisic acid	ABA	264.3
Gibberellic acid	GA3	346.4

Professor Thimann was a pioneer who played a major role in plant physiology by providing detailed descriptions and functions of hormones in the control and development of plants. The discovery of IAA was of seminal importance in plant tissue culture and the agricultural/horticultural industry as a whole (Stowe 1999). When Hinchee et al. (1988) prepared cotyledon explants from soybean seedlings, Gamborg's B_5 medium containing 1.15 mg/l benzyladenine (BAP) was used. Prepared cotyledonary explants were subcultured for shoot induction on a fresh B_5BAP medium containing 500 mg/l carbenicillin and 100 mg/l cefotaxime, together with 200–300 mg/l kanamycin.

The report clearly indicates that the production of transgenic soybean plants was made possible by both the procedure that utilised *Agrobacterium*-mediated DNA transfer and the B_5BAP culture medium composition. This was in addition to other several transformation parameters such as the soybean cultivar and antibiotic selection regime used. Apart from this report, Li et al. (2017) optimised *Agrobacterium*-mediated transformation by improving the infection efficiency of this bacterium and regeneration efficiency of cotyledonary explants in soybean cultivars Jack Purple and Tianlong.

These cultivars, respectively, recorded 7 and 10% transformation efficiency when B5 basal culture medium supplemented with 1.67 mg/l BAP was used for shoot induction, and 0.5 mg/l GA_3 and 0.1 mg/l IAA as well as 1 mg/l Zea were used on MS medium for the elongation of induced shoots. These plant tissue culture applications demonstrate a crucial role that this regeneration system plays in facilitating the delivery of exogenous DNA into host plants. Furthermore, over 85% of transgenic plants were obtained using the *Agrobacterium*-mediated genetic transformation that predominantly used plant tissue culture as a prerequisite regeneration method (Li et al. 2017). Tissue culture steps included in the transformation process include the following:

1) Obtaining seedlings used as explant source from germination medium;
2) Transferring *Agrobacterium*-infected explants onto a cytokinin culture medium for shoot induction;
3) Elongating the induced shoots on elongation medium supplemented with auxins, and often modified further with antibiotics used as selectable markers;
4) Rooting of the elongated shoots on auxin containing culture medium.

Then, after *in vitro* culture of explants, regenerated plants were transferred to pots, acclimatised and grown to maturity, until flowering as well as seed production stages. During this process, many factors such as *Agrobacterium* strain, plant genotype/cultivar, selection medium, infection time, co-cultivation period and medium composition, etc., ultimately influence transformation efficiency. The effect may either be positive or negative depending on the state and conditions of optimisation. Soybean transformation has in many ways demonstrated its own limitations like low rates of transgenic plants recovery, genotype specificity and the lack of a routine protocol suitable for wider applications on varieties of this recalcitrant crop.

2.7 SUMMARY

The literature shows that studies that initiated plant transformation first came from Chilton et al. (1980), Otten et al. (1981), Ooms et al. (1981) and then a breakthrough by Luis Herrera-Estrella and his colleagues in 1983. But despite these later modifications, earlier publications also indicated that the first signs of genetic transformation came from a report by Schilperoort and colleagues around 1967. These researchers synthesised a short RNA strand from a complementary *Agrobacterium* DNA template isolated from a cultured *Nicotiana* tumour which then indicated that bacterial DNA had been indeed transferred into plant cell (Somssich 2019). However, Chilton et al. (1977) presented solid evidence indicating the incorporation of a bacterial DNA segment in the plant tumour cells. Separating tumour DNA into genetic fragments after endonuclease SmaI restriction enzymes explicitly indicated that T-DNA was incorporated into tumourous tissues of the plant host.

This evidence led to many transformation hypotheses, including the following: (1) tumour formation principle, (2) replacement of genes in the T-DNA region, (3) insertion of genes of interest and (4) disarming the Ti-plasmids to obtain non-oncogenic plasmid variants/binary vectors that are currently utilised. All of the above, in addition to the discovery of plant tissue culture by Gottlieb Haberlandt and plant hormones by Kenneth V. Thimann, opened avenues and brought about the tissue culture-based genetic engineering technology that revolutionised modern breeding in plants. Even more impressively, it now became clear that *Agrobacterium tumefaciens* did not only accelerate the expression of foreign DNA molecules in different hosts but remains the simplest and most affordable approach used in the genetic manipulation of all living organisms/cells of plants, fungi, yeast or humans.

Furthermore, these wider applications demonstrate the impact that this biotechnological technique may continue to have in the improvement of genetic diversity of many crop species conferring tolerance of various stress factors. Thus, considering the main current limitations and potential of genetic transformation in plants, especially soybean crop, it can continue to be used for improving crop performance against the ever-changing climatic conditions that frequently bring newly evolved insect pests, drought and severe weather conditions. There is ample information and techniques used for crop improvement, but history and current research show that the *Agrobacterium tumefaciens*-mediated genetic transformation approach presents numerous agroeconomic advantages over conventional breeding. These include the ability to transfer and express genes across genetically diverse organisms, high heritability or gene transmission in a Mendelian fashion, less complex/low copy number, short breeding periods and rapid registration of new cultivars.

2.8 ABBREVIATIONS

CaMV	Cauliflower mosaic virus
Chv	Chromosomal virulence
Da	Daltons
DNA	Deoxyribonucleic acid
Goi	Gene of interest
GUS	β-glucuronidase
PAT	Phosphinothricin transferase
PGRs	Plant growth regulators
T-DNA	Transferred DNA
Ti-plasmid	Tumour-inducing plasmid
RNA	Ribonucleic acid
Vir	Virulence

REFERENCES

Basiran N, Armitage P, Scott RJ and Draper J. (1987). Genetic transformation of flax (*Linum usitatissimum*) by *Agrobacterium tumefaciens*: Regeneration of transformed shoots via callus phase. *Plant Cell Reports* 6, 396–399.

Bevan M. (1984). Binary *Agrobacterium* vectors for plant transformation. *Nucleic Acids Research* 12(22), 8711–8721.

Bevan MW, Flavell RB and Chilton MD. (1983). A chimaeric antibiotic resistance gene as a selectable marker for plant cell transformation. *Nature* 304, 184–187.

Campbell CL. (1983). Erwin Frink Smith- Pioneer plant pathologist. *Annual Review in Phytopathology* 21, 21–27.

Chilton MD, Drummond MH, Merlo DJ, Sciaky D, Montoya AL, Gordon MP, and Nester EW. (1977). Stable incorporation of plasmid DNA into higher plant cells: The molecular basis of crown gall tumorigenesis. *Cell* 11(2), 263–271.

Chilton MD, Saiki RK, Yadav N, Gordon MP and Quetier F. (1980). T-DNA from *Agrobacterium* Ti plasmid is in the nuclear DNA fraction of crown gall tumor cells. *Proceedings of the National Academy of Science of the United States of America* 77, 4060–4064.

De Block M, Herrera-Estrella L, Montagu MV, Schell J and Zambryski PC. (1984). Expression of foreign genes in regenerated plants and in their progeny. *EMBO Journal* 3(8), 1681–1689.

Escudero J, Neuhaus G and Hohn B. (1995). Intracellular *Agrobacterium* can transfer DNA to the cell nucleus of the host plant. *Proceedings of the National Academy of Science of the United States of America* 92, 230–234.

Fry J, Barnason A and Horsch RB. (1987). Transformation of *Brassica napus* with *Agrobacterium tumefaciens* based vectors. *Plant Cell Reports* 6, 321–325.

Garfinkel DJ, Simpson RB, Ream LW, White FF, Gordon MP and Nester EW. (1981). Genetic analysis of crown gall: Fine structure map of the T-DNA by site-directed mutagenesis. *Cell* 27, 143–153.

Gelvin SB. (2012). Traversing the cell: Agrobacterium T-DNA's journey to the host genome. *Frontiers in Plant Science* 3, 52, 1–11.

Herouet C, Esdaile DJ, Mallyon BA, Debruyne E, Schulz A, Currier T, Hendrickx K, van der Klis R-J and Ruan D. (2005). Safety evaluation of the phosphinothricin acetyltransferase proteins encoded by the *pat* and *bar* sequences that confer tolerance to glufosinate-ammonium herbicide in transgenic plants. *Regulatory Toxicology and Pharmacology* 41(2), 134–149.

Herrera-Estrella L. (2020). My journey into the birth of plant transgenesis and its impact on modern plant biology. *Biography of Pioneers in Plant Biotechnology* 18(7), 1487–1491.

Herrara-Estrella L, Depicker A, Montagu MV and Schell J. (1983). Expression of chimaeric genes transferred into plant cells using a Ti-plasmid-derived vector. *Nature* 303, 209–213.

Hinchee MAW, Connor-Ward DV, Newell CA, McDonnell RE, Sato SJ, Gasser CS, Fischhoff DA, Re DB, Fraley RT and Horsch RB. (1988). Production of transgenic soybean plants using *Agrobacterium*-mediated DNA transfer. *Biotechnology* 6, 915–921.

Hwang H-H, Wu ET, Liu S-Y, Chang S-C, Tzeng K-C and Kado I. (2013). Characterisation and host range of five tumorigenic *Agrobacterium tumefaciens* strains and possible application in plant transformation assays. *Plant Pathology* 62(6), 1384–1396.

Hwang H-H, Yu M and Lai E-M. (2017). *Agrobacterium*-mediated plant transformation: Biology and applications. *Arabidopsis Book* 15, e0186, 1–31.

Jung SM, Hur YY, Preece JE, Fiehn O and Kim YH. (2016). Profiling of disease-related metabolites in grapevine internode tissue infected with *Agrobacterium vitis*. *The Plant Pathology Journal* 32(6), 489–499.

Kerr A. (1969). Transfer of virulence between isolates of *Agrobacterium*. *Nature* 223, 1175–1176.

Koncz C, De Greve H, Andre D, Deboeck F, Montagu MV and Schell J. (1983). The opine synthase genes carried by Ti plasmids contain all signals necessary for expression in plants. *EMBO Journal* 2, 1597–1603.

Krikorian AD and Berguam DL. (2003). Plant cell and tissue cultures: The role of Haberlandt. In Laimer M and Rucker W (eds). *Plant Tissue Culture*. Springer, Vienna. pp. 25–53.

Li S, Cong Y, Liu Y, Wang T, Shuai Q, Chen N, Gai J and Li Y. (2017). Optimisation of *Agrobacterium*-mediated transformation in soybean. *Frontiers in Plant Science* 8(246), 1–15.

Mangena P. (2021). Synthetic seeds and their role in agriculture: Status and progress in sub-Saharan Africa. *Plant Science Today* 8(2), 1–8.

McCormick S, Niedermeyer J, Fry J, Barnason A, Horsch R and Fraley R. (1986). Leaf disc transformation of cultivated tomato (*L. esculentum*) using *Agrobacterium tumefaciens. Plant Cell Reports* 5, 81–84.

Nester EW. (2015). *Agrobacterium*: Nature's genetic engineer. *Frontiers in Plant Science* 5, 730, 1–16.

Ooms G, Hooykaas PJJ, Moolenaar G and Schilperoort RA. (1981). Grown gall plant tumors of abnormal morphology, induced by *Agrobacterium tumefaciens* carrying mutated octopine Ti plasmds; analysis of T-DNA functions. *Gene* 14, 33–50.

Otten L, Greve HD, Hernalsteens J-P, Montagu MV, Schieder O, Straub J and Schell J. (1981). Mendelian transmission of genes introduced into plants by the Ti plasmids of *Agrobacterium tumefaciens. Molecular and General Genetics* 183, 209–213.

Paz MM, Martinez JC, Kalvig AB, Fonger TM and Wang K. (2006). Improved cotyledonary-node method using an alternative explant derived from mature seed for efficient Agrobacterium-mediated soybean transformation. *Plant Cell Reports* 25, 206–213.

Poncet C, Antonini C, Bettachini A, Hericher D, Pionnat S, Simonini L, Dessaux Y and Nesme X. (1995). Impact of the crown gall disease on vigour and yield of rose trees. *Acta Horticulurae* 424, 221–225.

Shokraii EH and Azizian D. (1979). A study of the tumorigenesis of *Agrobacterium tumefaciens* Conn. on some plant species of Iran. *Zentralbl Backteriol Naturwiss* 134(4), 335–342.

Smith EF and Townsend CO. (1907). A plant-tumor of bacterial origin. *Science* 25, 671–673.

Somssich M. (2019). A short history of plant transformation. *Peer Journal Reprints* 2(27556), 1–28.

Stowe BB. (1999). Kenneth V. Thimann (5 August 1904–15 January 1997). *Proceedings of the American Philosophical Society* 143(3), 502–509.

Trigiano RN and Gray DJ. (2005). *Plant Development and Biotechnology.* CRC Press, Boca Raton, pp. 9–14.

Umbeck P, Johnson G, Barton K and Swain W. (1987). Genetically transformed cotton (*Gossypium hirsutum* L.) plants. *Bio/Technology* 5, 263–266.

Xu N, Yang Q, Yang X, Wang M and Guo M. (2020). Reconstruction and analysis of a genome-scale metabolic model for *Agrobacterium tumefaciens. Molecular Plant Pathology* 22, 348–360.

Zaenen I, van Larebeke N, Montagu MV and Schell J. (1974). Supercoiled circular DNA in crown-gall inducing *Agrobacterium* strains. *Journal of Molecular Biology* 86, 109–127.

Zambryski PC, Holsters M, Kruger K, Depicker A, Schell J, Montagu MV and Goodman HM. (1980). Tumor DNA structure in plant cells transformed by *A. tumefaciens. Science* 209, 1385–1391.

3 *Agrobacterium tumefaciens*

3.1 INTRODUCTION

The name *Agrobacterium tumefaciens* does not merely refer to just a common plant pathogenic microorganism within prokaryotic eubacteria, but a valuable nature's genetic engineer that plant molecular biologists cannot appreciate enough. More than two decades ago, the idea of using this bacterial species as a vector to create transgenic plants was only viewed as a prospect. But currently, some monocotyledonous and dicotyledonous plant species are routinely transformed using this bacterium as a biological vector (Gelvin 2003). This soil-borne pathogenic Gram-negative bacteria has the unique capability to transfer its portion of the genome, called the transferred DNA (T-DNA), into foreign cells and change the genetic makeup of such recipient host cells. When comparing species found in the Rhizobiaceae family, especially the genus *Agrobacterium*, *A. tumefaciens* serves as the most important and best strain used in the genetic engineering of crop plants.

Agrobacterium radiobacter (an avirulent species), *Agrobacterium rhizogenes*, *Agrobacterium rubi* and the recently introduced species *Agrobacterium vitis* form a group of agronomically important species of bacteria found within this genus. These motile, rod-shaped soil-borne bacterium do not produce spores, but they reproduce by binary fission and are closely associated with nitrogen (N) fixing species of *Rhizobium* bacteria which form root nodules in leguminous plants (Pitzschke 2013).

3.2 CLASSIFICATION

Agrobacterium tumefaciens forms part of a diverse group of alphaproteobacterium in the family Rhizobiaceae. Its genus *Agrobacterium* is composed of a small number of species that are classified mostly on the basis of disease symptomology and host range (Gelvin 2003). As we have seen with other species within this genera, *A. tumefaciens* has not been validly published under the bacteriological codes of rules of the International Code of Nomenclature. However, the bacterium retains the status of a type species of the genus *Agrobacterium* due to its greater prevalence in scientific literature as shown on Table 3.1 (Schoch et al. 2020). Young et al. (2001) recently reclassified this species as *Rhizobium radiobacter*. This classification sparked a fierce ongoing international debate due to the fact that very little morphological and pathological data were considered as compared to sufficient molecular data for supporting this assertion.

This species was reclassified as *Rhizobium radiobacter*, originally from *A. tumefaciens* (Smith and Townsend 1907) Conn 1942 by analysing 16S rDNA sequence

DOI: 10.1201/b22829-3

TABLE 3.1
Systematics information of *Agrobacterium tumefaciens* compiled using information from the National Centre for Biotechnology Information (NCBI)

Domain/kingdom	**Bacteria**
Phylum	Proteobacteria
Class	Alphaproteobacteria
Order	Rhizobiales
Family	Rhizobiaceae
Genus	*Agrobacterium*
Species/scientific name	*Agrobacterium tumefaciens*

Synonyms of this bacterium such as *Alcaligens radiobacter*, *Achromobacter radiobacter*, *Bacillus radiobacter*, *Bacterium radiobacter*, *Rhizobium radiobacter* and *Psedomonas radiobacter* are validly recognised in the classification since their publication by Beijerinck and van Delden in 1962 (Horst 1983; Schoch et al. 2020).

data and some phenotypic generic circumscriptions between *Agrobacterium*, *Allorhizobium* and *Rhizobium*. However, as it currently stands, *A. tumefaciens* is classified as a strain of the genus *Agrobacterium* capable of inducing tumourigenic reactions in a wide range of host plant species as envisaged earlier by Smith and Townend (1907). A major setback is, however, the fact that many microbial species do not have fossil records available to establish their evolutionary history. Some of the pitfalls relating to their fossilisation include problems in translating data acquired from such microbes into robust fossils, determining an exact length of time for fossilisation and inconsistent outcomes on experimental methods used for bacterial preservation to approximate the original bacterial form (Shen et al. 2018).

In the past, morphological and physiological characteristics were the major parameters used for classifications, but currently, genomic and proteomic data in combination with metabolomics also play a crucial role in determining species lineage and the exact species identity. Other information like phenotypic and metabolic data, although useful, is easily influenced by environmental factors which make classification very difficult and frustrating. *A. tumefaciens* remains grouped together in the same genus with *A. rhizogenes* capable of causing hairy-root disease in hosts, and *A. vitis* causing gall-disease in grapevines which controversially proved to be highly specific to this Vitaceae family.

The genus *Vitis* contains about 79 accepted species, of which most of the vines are used for direct consumption as fruit and through fermentation to produce wine beverages. The majority of grape cultivars used for human consumption fall within the species *Vitis vinifera*, and this genera consists of shrubs and woody lianas that climb using leaf tendrils (Keller 2015). Plant specificity between *A. vitis* and *Vitis* genotypes served as one of the points of contention by Young et al. (2001), further indicating that this relationship required an alternative classification from the *Agrobacterium* genus. These species are still within the *Agrobacterium* genus as Gram-negative bacteria established by Harold Joel Conn, an American agricultural bacteriologist who classified them as bacterial species causing horizontal gene transfer that induce tumours in plants (Stonier 1959).

3.3 GENERAL CHARACTERISTICS OF *AGROBACTERIUM*

All bacteria are unicellular organisms that have some common notable traits such as a small, microscopic size and lack of membrane-bound organelles and have their genetic chromatin material scattered throughout the cytoplasm. The kinds of membrane-bound organelles that are lacking inside bacterial cells include the nucleus, endoplasmic reticulum, Golgi apparatus, mitochondria, plastids, lysosomes and vacuoles predominantly found in multicellular eukaryotic cells. Under ideal conditions, the growth activities of bacteria, including *Agrobacterium*, proceeds in several stages termed the *Lag*, *Log*, *Stationary* and *Death* phases (see Figure 3.1). The growth curve is a representative of the logarithmic growth of a number of typical living bacterial cells plotted as a function of time. Even though the logarithmic growth of bacteria is well established, data are very scant on the mechanism and dynamics of growth of *Agrobacterium tumefaciens* in soils, rhizosphere or other solid substrata.

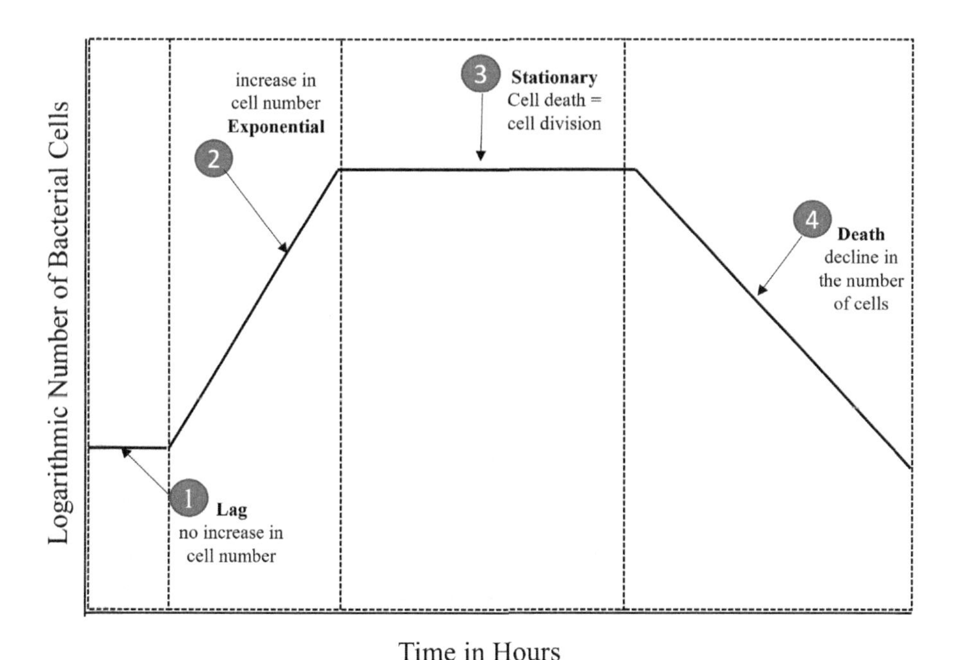

FIGURE 3.1 A typical bacterial growth curve established through *in vitro* culture of bacterial cells on a nutrient medium.

Current difficulty experienced on growth analysis of bacteria around this 'microbial storehouse' is caused by the biological and chemical features of soils, together with the influence from plant roots. Furthermore, the rhizosphere harbours a large bacterial population diversity responsible for some of the key activities in biogeochemical cycling of organic matter and mineral nutrients (Lagos et al. 2015). Bias in classical molecular identification techniques and the escalating costs of next-generation sequencing platforms and post-genomic techniques constitute some of the major limitations that make studying bacteria in the rhizosphere challenging. But the mechanism that *Agrobacterium* uses to infect plant tissues and integrate a portion of its DNA into host genomes resulting in tumours and plant metabolic changes have been comprehensively studied for over a century.

Apart from the processing and transfer of a specific T-DNA portion from the *Agrobacterium*'s tumour-inducing (Ti) plasmid as indicated in the previous chapter, the formation of crown galls by wounded site infections in dicotyledonous and some monocotyledonous plants is made possible by the involvement of other bacterial cell characteristics. Many of these characteristics, which are mostly morphological, include anatomical features of most Gram-negative bacteria like *Agrobacterium* cell wall presented in Figure 3.2. However, it is important to note that these characteristics are complemented by the bacterium's excellent and precise molecular mechanism of infection, which is not part of the main focus in this section. *Agrobacterium*

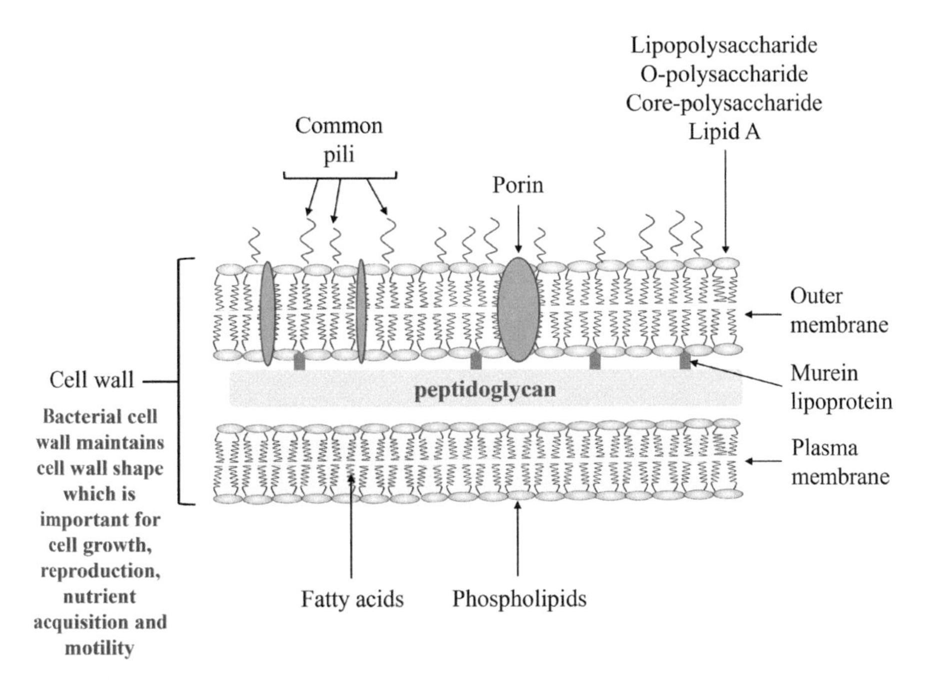

FIGURE 3.2 Gram-negative cell wall structure showing the unique murein (peptidoglycan) layer that is mainly found in bacteria. Peptidoglycans have not been found anywhere else on Earth other than the bacterial cell wall.

species are rod-shaped and motile, having one to six peritrichous flagella. These bacterial cells can exist in pairs or as single cells ranging between 0.6 and 1.0 μm in length with diameters between 1.5 and 3.0 μm (Horst 1983).

Rod-shaped bacterial cells are normally called bacilli (singular, bacillus) referring to morphological differences but not so much taxonomical similarities in this case. The ends of rods also vary in shape and may be bifurcated, flat or round shaped. The formation of pili (Figure 3.2) by *Agrobacterium* also requires the induction of tumour-inducing plasmid encoded virulence genes, and its formation is severely affected by low temperatures (Fullner et al. 1996). In many bacteria these structures are required for interbacterial DNA transfers. Fullner and colleagues demonstrated through genetic analysis that *VirA*, *VirB1–VirB11* and *VirD4* are the only virulence Ti-plasmid genes required for pili assembly. Furthermore, this study demonstrated that a loss and gain of pili are also correlated with effective gains of T-DNA transfer in host cells. This clearly showed that pili are important for DNA transfers into host plants, a similar mechanism to conjugation between bacterial cells. Large filaments (flagella) that are circumthecally arranged near one end of the bacilliform cells are also required for motility.

The bacteria chemotactically move into wounded plant tissue using these structures, and they possibly assist in the orientation of bacterial cells at the various sites of infection as well. According to Chesnokova et al. (1997) flagella are required by the bacteria to facilitate motility reaching root surfaces and are encoded by the *FlaA*, *FlaB* and *FlaC* genes facilitating virulence. *Agrobacterium* cell walls, on the other hand, protect the cell and determine to a great extent its shape. The cell wall also contains a polysaccharide component called the peptidoglycan layer shown in Figure 3.2. All Gram-negative and Gram-positive bacteria contain varied cell wall components, including the higher concentrations of lipid bilayer playing an important role in wall porosity and the ultimate classification of bacteria using the Gram staining technique developed by Hans Christian Gram (Agrios 1988).

3.4 SIZE AND MORPHOLOGY

Like many other bacteria, *Agrobacterium* spp. are characteristically smaller than even microscopic eukaryotic organisms. Bacteria range in size from 0.2 to 1.5 μm in width and 1 to 6 μm in length. The differences in the size of bacteria versus eukaryotic cells are considerably significant. As indicated in the above section, *Agrobacterium tumefaciens* is about 0.6–1.0 × 1.5–3.0 μm in size. Bacterial species in the Alphaproteobacteria are fairly large compared to species of Gram-positive bacteria, Spirochetes, and even to some genera within Proteobacteria such as Gammaproteobacteria and Cytophaga-like bacteria. The Spirochetes, for instance, contain very thin bacteria with species of about 0.1 to 0.15 μm in diameter, with some of the species within this genus being about 5 to 6 μm in length, fairly comparable to some species of *Agrobacterium* (Cotterell and Kirchman 2004). Generally speaking, the structure and morphology of bacteria are important for the metabolic processing required for the microorganism's survival and reproduction.

The rate of microbial metabolism has been proved to increase with an increase in the surface/volume ratio of the cells. When comparing the phyla or domains, this ratio appeared to be about ten times greater in prokaryotic cells than eukaryotic organisms. A small size has major consequences in dictating some biological properties. For instance, the high surface-volume ratio further implies that cytoplasmic contents are more concentrated and in close and free contact with each other. Furthermore, the lack of internal compartmentalisation also results in a free movement of ions and molecules within the cells. These movements are easily translated into accelerated metabolism, rapid growth and cell division (Sonea and Mathieu 2018). As will be discussed in subsequent sections below, cells of *Agrobacterium* species also contain large tumour-inducing (Ti) plasmid replicons to perform essential functions such as growth and reproduction. Therefore, *A. tumefaciens* with its proportionately large surface/volume area can effect rapid metabolic exchanges, protein complex and genetic material transfers across cells found within the surrounding environment.

Agrobacterium as a prokaryotic cell does not only constitute a larger surface-volume ratio but is also complemented by its morphology. The combined cell characteristics like the flagella/pilli, porins, cell walls and their cellular genetic components (circular chromosomes, linear chromosomes and plasmids) of both molecular and morphological nature play a critical role during pathogenesis. Finally, cell size, structure and morphology have direct consequences for the ecology, adaptability and evolution of all prokaryotic bacteria, including *A. tumefaciens* (Smith and Townsend 1907; Agrios 1988).

3.5 NUTRITION

Agrobacterium species do not obtain energy and nutrients by performing photosynthesis. Many species of bacteria like the Cyanobacteria, green-sulphur bacteria, purple bacteria and Prochlorophytes are capable of obtaining energy for growth and metabolism from organic acids or carbon monoxide using light to evolve hydrogen (Idi et al. 2014). Photosynthetic bacteria have long been studied for their capacity to produce the hydrogen used as an electron donor during photosynthesis. Like in plants, hydrogen production is used to establish the proton motive force or hydrogen proton gradient used in the synthesis of reducing powers and adenosine triphosphates (ATP) by ATP-synthase enzyme complex. Photosynthetic bacteria contain reaction centres noncovalently bound to integral membrane proteins, and typically possess a strong near-infrared absorption band between 800 and 890 nm (Cogdell et al. 2020). However, *Agrobacterium tumefaciens* mostly exist as auxotrophs, a microorganism that is unable to synthesise essential nutrients, often due to gene mutations.

This bacterium acquires its nutrients from the hosts through the initiation of tumours or crown-gall diseases. Such nutritional defects also reflect the existing differences in pathogenicity and host range specificity found within the entire genus *Agrobacterium* (Lippincott and Lippincott 1969). Many bacterial strains from this group fail to induce tumours because of nutritional limitations, especially if their ability for host infection is reduced or inhibited. As indicated by Lippincott, this implies that the ability of the bacteria to cause crown-gall tumours can depend upon the interplay between the bacterium and nutritional qualities of plant hosts' wound sites. Studies on the nutritional requirements of *A. tumefaciens* have so far provided major insights on the pathogenicity of this bacterium and led to the improvement of the intended genetic transformation in most crop plants using such information. In general, species of *Agrobacterium* obtain their energy from the aerobic oxidation of organic compounds and do not carry out anaerobic metabolism or fermentation.

In addition, these strains feed themselves by using and controlling plant systems as their nutritional machinery to regulate and produce unique nutrients called opines. Opines and pseudo-opines genes encoded by strains of *Agrobacterium* are utilised by the bacteria for nutrient acquisition during host colonization. Opines serve as a source of carbon, nitrogen, phosphorus and sulphur used for the synthesis of amino acids and other metabolites (Flores-Mireles et al. 2012). Further evidence showed that different strains of *Agrobacterium* transfer diverse types of opine biosynthetic genes through their tumour-inducing plasmids. An example of this is the so-called octopine type Ti-plasmids such as pTiAch5, pTiA6, pTiB6 and pTiR10 that transfer *mas2*, *ocs*, *ags* and *mas1* opine biosynthetic genes, respectively (Hack and Kemp 1980; Firmin et al. 1985). These genes, including discrete fragments of the Ti-plasmid oncogenic DNA that are expressed in plant nuclei, are important for the production of phytohormones and nutrients used during bacterial colonisation in host tissues.

3.6 REPRODUCTION

The most common method of bacterial reproduction is binary fission. The advanced form of reproduction, sexual reproduction, is absent in bacteria. Binary fission is characterised by an approximately equal separation of the bacterial cell into two daughter cells that are often differentiated into either transversely or longitudinally separated cells (Angert 2005). Similar to other species of eubacteria, the mode of asexual reproduction is also used by species of Rhizobiaceae (*Rhizobium* and *Agrobacterium*) for cell separation along the transverse axis. Not all prokaryotes use binary fission for reproduction, but multiple offspring formation and budding could also be used. Division of the cells is then followed by cell wall synthesis. In some bacterial strains, like *Agrobacterium tumefaciens*, cell multiplications are greatly influenced by nutrient availability on the wound site as indicated in the previous section above.

Brisset et al. (1991) found that the bacterial population curves for *Agrobacterium* strains at the wound sites were greatly influenced by the nutritional profile. Polygalacturonase (PG) enzyme production and PG metabolism on this site affected multiplication rates of *A. tumefaciens*. Polygalacturonase enzymes are classified into the class of glycosyl hydrolases that functions to catalyse the hydrolysis of α-1,4-glycosidic bonds of polygalacturonase pectic acids. Polygalacturonase is chromosomally encoded and associated with the attachment of the bacterium to host cell surfaces during the infection process apart from cell multiplication. Brisset et al. (1991) reported that only the polygalacturonase-producing strains of *A. tumefaciens*, biovar CG39 and CG484, were able to multiply exponentially for about 100 hours at the wound site on grape roots. Furthermore, this logarithmic transverse binary fission caused root necrotic symptoms which ultimately became visible within 48 hours of inoculation.

Similar to *Rhizobium*, *Agrobacterium* species rely on binary fission for their reproduction and do not use other alternative means. Methods such as budding are relatively less common within this genus. But budding is commonly used in *Caulobacter*, *Hydromicrobium* and *Stella* species. Vollmer et al. (2008) found cell division through binary fission to be highly conserved and regulated based on the interaction between filamenting temperature-sensitive mutant Z (FtsZ) protein and the peptidoglycan biosynthesis machinery. FtsZ is considered to be an almost universally conserved bacterial cytoskeletal element constituted in the binary fission process (Rivas-Marin et al. 2020), while budding is considered a reproductive process only used if the divided cells are smaller than parent cells and when the cytoplasmic content is newly synthesised (Vollmer et al. 2008).

3.7 CELL STRUCTURE

As briefly mentioned and illustrated in Figure 3.2, the cell wall critically functions to protect and isolate bacterial cytoplasmic materials from the surrounding environment. Most importantly, the cell wall of *Agrobacterium tumefaciens* protects it against detrimental environmental influences. This structure functions to provide cell rigidity, maintain cell shape, enable passage for the exchange of molecules into or out of the cell and allow the bacterial cell to keep its high internal osmotic pressure (Sonea and Mathieu 2018). *A. tumefaciens* as a biotroph also uses the cell wall or specialised structure on the cell wall to establish a long-feeding relationship with living cells of the plant host. However, cell wall structures like pili (singular, pilus) are considered to be generally required for interbacterial DNA transfers. More evidence points out that pilus assembly is necessary for both the injection of DNA into recipient host cells and mediation of contact formation between the bacteria and its hosts (Fullner et al. 1996; Baron and Zambryski 1996).

However, some reports, like those of Baron and Zambryski (1996), earlier suggested that DNA transfer channels are formed from other components of transfer apparatus associated with the cell wall. More evidence indicated that adhesive pili also proved to be associated with virulence factors of the transferred plasmid DNA of various pathogenic bacteria. Although not yet assigned a role in pathogenesis and tumour formation, many scientists still believe that the pilus structure determines *A. tumefaciens* pathogenicity. Lai et al. (2002) reported that pili are generated when *Agrobacterium* cells are induced naturally through host plant phenolic compounds from the wound site or experimentally by the additive of an inducer, acetosyringone. Acetosyringone, also known as 3',5'-dimethoxy-4'-hydroxy-acetophenone, is a phenolic natural product used in plant tissue culture involving *A. tumefaciens*.

This chemical functions to attract the bacteria to the wound site and induce virulence genes to facilitate the transfer of T-DNA region across cell walls. The induction of these phenols is said to specifically lead to virulence gene expressions involved in the synthesis and assembly of the VirB2 protein (propilin) forming the pilus. Even though *Agrobacterium* T-pili attach to the cell wall surfaces, these fimbriae remain protein structures that are an extension from the cell wall envelope. In contrast to Gram-negative bacteria, Gram-positive microbes assemble the cell wall-surface multisubunit protein polymers using sortase. This enzyme joins individual pilin monomers to form covalent polymer in the cell wall peptidoglycan. It is the role of sortases, as a group of prokaryotic enzymes, to modify surface proteins by recognising and cleaving a carboxylic terminal sorting signal.

The pili assembly mechanism is also used by Gram-positive bacteria for the colonisation of specific host tissues, the modulation of host immune responses and the development of bacterial biofilms (Mandlik et al. 2008). In both groups of bacteria, sugar derivatives such as N-acetylglucosamine and N-acetylmuramic acid cross-linked with repeating network structure of amino acids that form peptidoglycan are responsible for cell wall strengthening (see Figure 3.2 structure for the positioning of peptidoglycan). Approximately 80 to nearly 90% of both Gram-positive and Gram-negative bacterial structural compositions in general are made up of peptidoglycan. *A. tumefaciens*, however, consist of a single layer of peptidoglycan surrounded by the outer and inner membranous structures. Finally, the surface composition of a bacterium enables the organism to sense and assess its environment (Mandlik et al. 2008) and then function to induce the growth of tumour-like structures on host species.

3.8 PROTOPLASMIC MATERIALS

The genetic information for tumours or crown-gall formation is encoded on the Ti-plasmid found in bacterial cytoplasm. This is the portion of the Ti-plasmid (the T-DNA) that *A. tumefaciens* make use of its virulence factors and proteins to colonise host tissues to transport DNA into the nucleus of transformed plants. *Agrobacterium* contains a granular cytoplasm with one or more rounded, dense inclusions called chondroids. Rubio-Huertos and Desjardins (1960) associated this low-density core of granular protoplasm with its nuclear structures. Generally, this gel-like matrix of the bacterial cytoplasm contains 80% water, enzymes, inorganic ions, many low molecular weight compounds, ribosomes, chromatin DNA materials and plasmids. Amongst these, plasmids are of greater interest to scientists who have for decades studied their genetics and physiological and morphological responses, especially their symbiotic and pathogenic mechanisms on plant hosts.

All activities in the cell, including the expression of bacterial virulence (*vir*) genes for decoding of *vir* proteins and synthesis of the single-stranded T-DNA molecules from the Ti-plasmid, take place in the cytoplasm. The T-DNA associated *vir* proteins (T-complex) will be modified and injected into plant host cytoplasm to effect the translocation of this DNA and then the virulence proteins (VirD2, VirE2, VirE3, VirD5 and VirF) received by plant cells through a type IV secretion system (Pitzschke 2013). The bacterial type IV secretion system is a multisubunit that functions to transfer macromolecular proteins and nucleoproteins from bacterial cells across the membrane envelope to effect DNA transfer and expression inside host cells (Wallden et al. 2010).

Mediating the translocation of transcribed messenger RNA (mRNA) to proteins in the cytoplasm are ribosomes. Under the electron microscope, ribosomes appear randomly and equally distributed as aggregates and short linear arrays of black dots (Sonea and Mathieu 2018). They serve as an essential protein synthesis machinery that assembles amino acid ordered sequences from mRNA to make functional proteins that are already alluded to above. The characteristics mentioned above demonstrate the critical role that the cytoplasm plays in supporting a proportion of the hereditary information and plasmid DNA found within bacterial cells.

3.9 CHROMOSOMAL DNA AND PLASMIDS

Genetic transformation is largely known to be taking place in the Rhizobiaceae family, popularly associated with *Agrobacterium tumefaciens* and *Agrobacterium rhizogenes*. Species in the genus *Agrobacterium* contain genes responsible for the proper functioning of this mechanism organised in a circular chromosome, single linear chromosome and plasmid DNAs. According to Langley et al. (2019) a complete genome sequence of *A. tumefaciens* strain 33MFTa1.1 was predicted to have about 2,654 protein coding genes, 63 pseudogenes, 2 ribosomal RNA operons and 40 transfer RNAs and telomerase A (*telA*) gene encoding proteins required for hairpin loops at the end of linear chromosomes. About 1,800 protein coding genes, 69 pseudogenes, 2 ribosomal RNA (rRNA) operons and 14 rRNA were also predicted for the liner chromosome in contrast to the circular chromosome. Plasmid sequence containing candidate genes for plasmid replication initiation proteins (repA, repB and repC) and conjugative transfer (traA, traB, traC, traD, traF, traG, traH and traM) were also produced.

Generally, *Agrobacterium* species harbour one circular and one linear chromosomal DNA, in addition to the two plasmid DNAs containing the above genetic sequences for replication and conjugation. Furthermore, the circular and linear chromosomes get co-segregated and stably maintained within the bacterial cytoplasm as reported by Huang et al. (2012) in *A. tumefaciens* strain C58. *A. tumefaciens* has long been known to contain additional replicons in the form of circular plasmid DNAs. Virulent strains of this bacterium contain mega plasmids ranging from approximately 100 kilobase pairs to nearly 2 megabases that possess the capacity to induce tumours as indicated in previous discussions above (also see Chapter 8 for further details on plasmid DNAs). A second replicon is known as the binary vector, although this term also refers to the entire system that constitutes the two replicons (Komari et al. 2006).

The two replicons are separated by the fact that one carries the T-DNA containing the gene of interest and the other code for functions required for pathogenesis or virulence (Gordon and Christie 2014). As shown by Gordon and Christie (2014) most species of the class Alphaproteobacteria carry large plasmids whose functions rely upon coding elements, especially the repABC gene cassette for the replication, partitioning and maintenance of a few copy numbers per cell. These bacteria use plasmid DNAs to disseminate and maintain their genetic materials among populations of the class or genus, as well as to naturally engineer plants and non-plant species.

3.10 SUMMARY

Scientists have studied the use of *Agrobacterium tumefaciens* for the genetic improvement of crops and other organisms. However, studies during the last 20 years have shown that the processes involved in the transformation of plants are still lengthy, complicated and are accompanied by the fact that many genotypes, especially in the leguminous family, continue to show higher levels of inefficiencies or recalcitrance. These can be recognised as either the failure by researchers to efficiently optimise the transformation protocols in legume such as soybean, and encouragement for more research in culture optimisation, or for more avenues to be explored in the biology of *A. tumefaciens* in order to unravel the recalcitrance to *Agrobacterium*-mediated genetic manipulation of such crops. Plant transformation still poses numerous challenges despite ample information being available about its biology and mechanisms of pathogenesis to induce the formation of crow-gall tumours at wound sites of hosts by directly transforming plant cells.

As indicated by White and Winans (2007) the disease strategy befits the bacteria as it uses infected plant tissue as a machinery to produce opine nutrients for its proliferation and colonisation. Furthermore, many studies have demonstrated that all genes required for virulence, opine production and uptake/utilisation, including the expression of T-DNA, are carried within the Ti-plasmid DNAs or binary vector systems. These observations, made more than 20 years ago, demonstrated that the infection process thus relied upon multiple factors that include (1) the morphology and size of the bacterial cell or its surface/volume area, (2) the cell wall structure and largely on (3) the protoplasmic contents, which include ribosomes and hereditary information (Wang et al. 2015). All of these factors have a critical role to play, from the uptake of specific or free and soluble DNA molecules, to the incorporation of these exogenous fragments of DNA into recipient eukaryotic cells.

The well-organised cellular structure of *Agrobacterium* also ensures that all coded genetic sequences responsible for the proper functioning of this pathogenic mechanism are organised and contained inside competent replicons and operons. The result of this control and organisation is that the genes contained in these DNA functional units get effectively and efficiently expressed. The transforming replicons encode dozens of proteins or factors, some of which act as DNA receptor sites at the surface of host cells and then actively induce nutrient metabolism to ensure cell colonisation and DNA integration for subsequent disease formation.

3.11 ABBREVIATIONS

ATP Adenosine triphosphate
DNA Deoxyribonucleic acid
Fla Flagella
FtsZ Filamenting temperature sensitive mutant
N Nitrogen
NCBI National Centre for Biotechnology Information
PG Polygalacturonase

RNA	Ribonucleic acid
rRNA	Ribosomal RNA
T-DNA	Transferred DNA
Ti-plasmid	Tumour-inducing plasmid
tRNA	Transfer RNA
Vir	Virulence

REFERENCES

Agrios GN. (1988). *Plant Pathology*, 3rd eds. Academic Press Inc., London. pp. 558–565.

Angert ER. (2005). Alternatives to binary fission in bacteria. *Nature Review Microbiology* 3, 214–224.

Baron C and Zambryski PC. (1996). Plant transformation: A pilus in *Agrobacterium* T-DNA transfer. *Current Biology* 6(12), 15767–1569.

Brisset MN, Rodriquez-Palenzuela P, Burr TJ and Collmer A. (1991). Attachment, chemotaxis, and multiplication of *Agrobacterium tumefaciens* Biovar 1 and Biovar 3 on grapevine and pea. *Applied and Environmental Microbiology* 3, 214–224.

Chesnokova O, Coutinho JB, Khan IH, Mikhail MS and Kado CI. (1997). Characterization of flagella genes of *Agrobacterium tumefaciens*, and the effect of a bald strain on virulence. *Molecular Microbiology* 23(3), 579–590.

Christie PJ and Gordon JE. (2014). The *Agrobacterium* Ti plasmids. *Microbiology Spectrum* 2(6), 1–29.

Cogdell RJ, Isaac NW, Howard TD, McLuskey K, Fraser NJ and Prince SM. (2020). How photosynthetic bacteria harvest solar energy. *Journal of Bacteriology* 181(13), 3869–3879.

Cottrell MT and Kirchman DL. (2004). Single-cell analysis of bacterial growth, cell size and community structure in the Delaware estuary. *Aquatic Microbial Ecology* 34(2), 139–149.

Firmin JL, Stewart IM and Wilson KE (1985). N2–(1-carboxgethyl) methionine. A pseudo-opine in octopine-type crown-gall tumours. *Biochemistry Journal* 232(2), 431–434.

Flores-Mireles AL, Eberhard A and Winans SC. (2012). *Agrobacterium tumefaciens* can obtain sulphur from an opine that is synthesized by octopine synthase using S-methyl methionine as a substrate. *Molecular Microbiology* 84(5), 845–856.

Fullner KJ, Lara JC and Nester EW. (1996). Pilus assembly by *Agrobacterium* T-DNA transfer genes. *Science* 23(273), 1104–1109.

Gelvin SB. (2003). *Agrobacterium*-mediated plant transformation: The biology behind the “gene jockeying” tool. *Microbiollgy and Molecular Biology Reviews* 67(1), 16–37.

Hack E and Kemp JD. (1980). Purification and characterisation of the crown gall-specific enzyme, octopine synthase. *Plant Physiology* 65(5), 949–955.

Horst RK. (1983). *Compendium of Rose Diseases*. APS Press, St. Paul. pp. 23–25.

Huang WM, Da Gloria J, Fox H, Ruan Q, Tillou J, Shi K, Aihara H, Aron J and Casjens, S. (2012). Linear chromosome-generating system of *Agrobacterium tumefaciens* C58: Protelomerase generate and protects harpin ends. *Journal of Biological Chemistry* 287(30), 25551–25563.

Idi A, Md Mor MH, Wahab MFA and Ibrahim Z. (2014). Photosynthetic bacteria: An eco-friendly and cheap tool for bioremediation. *Review in Environmental Science and Bio/Technology* 14, 271–285.

Keller M. (2015). Botany and anatomy. In Keller M (eds), *The Science of Grapevines: Anatomy and Physiology*, 2nd eds. Academic Press, London. pp. 1–57.

Komari T, Takakura Y, Veki J, Kato N, Ishida Y and Hiei Y. (2006). Binary vectors and super-binary vectors. In Wang K (eds), *Agrobacterium Protocols*. Humana Press, Totowa, New Jersey. pp. 15–41.

Lagos L, Maruyama F, Nannipier P, Mora ML, Ogram A and Jorquera MA. (2015). Current overview on the study of bacteria in the rhizosphere by modern molecular techniques: A mini-review. *Journal of Soil Science and Plant Nutrition* 15(2), 504–523.

Lai E-M, Eisenbrandt R, Kalkum M, Lanka E and Kado CI. (2002). Biogenesis of T Pili in *Agrobacterium tumefaciens* requires precise VirB2 propilin cleavage and cyclization. *Journal of Bacteriology* 184(1), 327–330.

Langley S, Eng T, Wan KH, Herbert RA, Klein AP, Yoshikuni Y, Tringe SG, Brown JB, Celniker SE, Mortimer JC and Mulchopadhyay A. (2019). Complete genome sequence of *Agrobacterium* sp. strain 33MFTa1.1 isolated from *Thlaspi arvense* roots. *Genome Sequences* 8(37), e00432-19.

Lippincott JA and Lippincott BB. (1969). Tumour-initiating ability and nutrition in the genus *Agrobacterium*. *Journal of General Microbiology* 59, 57–75.

Mandlik A, Swierczynski A, Das A and Ton-That H. (2008). Pili in Gram-positive bacteria: Assembly, involvement in colonisation and biofilm development. *Trends in Microbiology* 16(1), 33–40.

Pitzschke A. (2013). *Agrobacterium* infection and plant defense- transformation success hangs by a thread. *Frontiers in Plant Science* 4(519), 1–12.

Rivas-Marin E, Peeters SH, Fernandez LC, Joglar C, Niftrik L, Wiegand S and Devos DP. (2020). Non-essentiality of canonical cell division genes in the planctomycete *Planctopirus limnophila*. *Scientific Reports* 10(66), 1–8.

Rubio-Huerto M and Desjardins PR. (1960). A vacuolar system in cells of *Agrobacterium tumefaciens* as shown by electron microscopy of ultra-thi sections. *Nature* 187, 1043–1044.

Schoch CL, Ciufo S, Domracher M, Hotton CL, Kannan S, Khovanskaya R, Leipe D, Mcveigh R, Oneill K, Robbertse B, Sharma S, Soussov V, Sullivan JP, Sun L, Turner S and Karsch-Mizrachi I. (2020). NCBI Taxonomy: A comprehensive update on curation, resources and tools. *Database, baaa062,* 1–21.

Shen B, Qiu J, Tenger B, Pan A, Yang Y and Bian L. (2018). Identification of bacterial fossils in marine source rocks in South China. *Acta Geochimica* 37(1), 68–79.

Smith EF and Townsend CO. (1907). A plant-tumor of bacterial origin. *Science* 25, 671–673.

Sonea S and Mathieu LG. (2018). *Prokaryology*. Presses de l'Universite de Montreal, Montreal, Canada. pp. 29–72.

Stonier T. (1959). *Agrobacterium tumefaciens* Conn. Production of an antibiotic substance. *Journal of Bacteriology* 79(6), 889–898.

Vollmer W, Blanot D and De Pedro MA. (2008). Peptidoglycan structure and architecture. *FEMS Microbiology and Reviews* 32, 149–167.

Wallden K, Rivera-Calzada A and Waksman G. (2010). Type IV secretion systems: Versatility and diversity in function. *Cellular Microbiology* 12(9), 1203–1212.

Wang L, Fan D, Chen W and Terentjev EM. (2015). Bacterial growth, detachment and cell size control on polyethylene terephthalate surface. *Scientific Reports* 5(15159), 1–11.

White CE and Winans SC. (2007). Cell-cell communication in the plant pathogen *Agrobacterium tumefaciens*. *Philosophical Transactions of the Royal Society B* 362(1483), 1135–1148.

Young JM, Kuykendall LD, Martincz-Romero E and Sawada AKH. (2001). A revision of Rhizobium Frank 1889, with an embedded description of the genus, and the inclusion of all species of *Agrobacterium Conn* 1942 and *Allorhizobium undicola de* Lajudie et al. 1998 as new combination: *Rhizobium radiobacter, R. rhizogenes, R. rubi, R. undicola* and *R. vitis*. *International Journal of Systematic and Evolutionary Microbiology* 51(1), 89–103.

4 Getting Started with *in Vitro* Soybean Transformation

4.1 INTRODUCTION

Plant transformation has emerged as an important genetic improvement tool for many monocots and dicot legumes such as soybean (*Glycine max* L. Merrill). This technology offers strategies for the overexpression or suppression of endogenous genes through the introduction of exogenous fragments of DNA in various host plant cells. The introduction of new genetic sequences, or manipulations of endogenous gene expressions via genetic transformation have the potential to give rise to new phenotypic variations. However, since the application of genetic transformation via *Agrobacterium tumefaciens* started, many studies surprisingly demonstrated that certain crops, particularly edible plants like soybeans, remain highly recalcitrant. Soybean serves as vital source of high-quality proteins and vegetable oils for both human and animal consumption. This crop consists of more than 36% protein, 30% carbohydrates, 19% oils and roughly 4% minerals and vitamins (Sharma et al. 2014).

But most researchers still sought to improve and optimise soybean plant transformation, at least for it to become a routine technique, by relying upon *Agrobacterium*'s natural ability to genetically engineer host cells as historically observed and described by the tumour-inducing principle. The successful development of transgenic plants using *A. tumefaciens*, coupled with plant tissue culture, requires a number of steps. The steps involved include the use of contamination-free seedling-derived explants for bacterial inoculation, co-cultivation of explants with *Agrobacterium* on a medium containing virulence inducers (for example, acetosyringone), efficient plant regeneration and recovery of transformed plantlets, as well as the positive identification of transgenics using reporter genes like the β -*glucuronidase* (*GUS*). These parameters are partly managed by optimising *in vitro* culture conditions to regulate the efficiency of gene transfer from bacterial plasmids and expression in explant cells of soybean for successful plant transformation.

Some cultures may require additional growth-promoting factors such as dithiothreitol, L-cysteine, PGRs and other organic additives (e.g., biotin, nicotinic acid or glutamate). A carbon source, sucrose or starch is readily used on tissue culture media (Gamborg's B5 medium and Murashige and Skoog medium) selected based upon the species to be cultured. There are several *in vivo* and *in vitro* protocols without the use of biological vectors and those involving *A. tumefaciens*. This chapter, however, describes a *A. tumefaciens*-mediated genetic transformation method using

DOI: 10.1201/b22829-4

seedling-derived explants obtained from soybean *in vitro* germinated seeds. The use of plant tissue culture for *Agrobacterium*-mediated transformation in soybean has been comprehensively established by Hinchee et al. (1988), reviewed and standardised by Zhang et al. (1999) and Paz et al. (2006).

Important strides in soybean transformation were made by these researchers, and their protocols are now optimised by many scientists around the globe. Such modifications include testing of the best decontamination methods and use of PGR-supplemented medium to produce stout seedlings serving as a source of explants. It is, thus, important to note that the costs involved in this transformation technique are very minimal, and no special arrangements/equipment are needed during the genetic improvement experiments. All of these factors, together with some of the most important stages involved in plant tissue culture-based *Agrobacterium tumefaciens*-mediated genetic transformation, will be reviewed in this chapter.

4.2 BACTERIAL STRAINS

Most strains of *Agrobacterium tumefaciens* and their avirulent relative strains of *A. radiobacter*, *A. rhizogenes* and *A. rubi* are continuously being explored for the genetic improvement of crops. These strains are easy and cheaper to handle under normal laboratory conditions, and their construct plasmids may remain viable for months under ultra-freezing conditions. Bacterial stocks are maintained in agar plates, or mostly stored and stabilised using glycerol stocks that prevent damage to the cell membranes and for keeping the cells alive for longer periods. As there are many methods to store bacterial cells, several of these have shown to be suitable for the handling and storage of *A. tumefaciens* and its relatives. To prepare this in a simplified manner, 0.5 mL of *Agrobacterium* liquid culture can be added into a sterile Eppendorf tube, followed by the addition of 0.5 mL of sterile 80% glycerol solution. The preservation of these cells is necessary, so that freezing does not change the morphological, physiological and genetic traits of the bacteria (Prakash et al. 2013).

The report by Prakash and colleagues consolidated the different methods used for the preservation of microorganisms with emphasis on cryopreservation and lyophilisation. Freeze drying, a lyophilisation method that is also known as cryodessication, is a process entailing water removal after cell freezing and then placing cells under vacuum pressure (Horn and Friess 2018). It would not probably serve a good purpose to use lyophilisation in *Agrobacterium* preservation, particularly because this bacterium can be stably and adequately preserved in liquid states. However, the use of low temperatures and a glycerol solution to cryopreserve *A. tumefaciens* may come with very minimal extrinsic and intrinsic factors affecting the survival and recovery of competent cells after long-term storage.

Generally, a good preservation method should minimise contamination, avoid inducing culture instability and promote cell viability to guarantee prolonged 'shelf-life' while reducing the risk of changing the genetic makeup of the organism. This includes genetic changes that may lead to the loss of the gene of interest (*goi*) contained within the T-DNA (van Asma 1995). Currently, a number of *A. tumefaciens* strains and plasmids are available for testing in the transformation of different plant

TABLE 4.1
Strains of *Agrobacterium tumefaciens* used for genetic improvement of plants

Strain	Plasmid
15955*	pTi1.5955
A6*	pTiA6
A208*	pTiT37-SE/ pMON9749
A856	pTiAg162
Ach5/ LBA4404*	pAL4404/ pBi121
Ag34	pAtAg34
B6*	pTiB6-806
Bo542*	pTiBo542
C58*	pTiC58/ pAtC58
CG54	pTiCG54/ pAtCG54
Chry5	pTiChry5
EHA101*	pTF101.1/ ΩPKY
R10*	pTiR10
T37*	pTiT37

Source: Knauf et al. 1983.
Strains with asterisk are well-known and widely used plasmids belonging to biotype 1 classification scheme.

and non-plant organisms as shown in Table 4.1, and are preserved in glycerol stocks at –80 to –86°C.

These strains, including their plasmids, are classified according to opines synthesised from the proteins expressed by T-DNA encoded genes (Morton and Fugua 2012). Among all of these plasmids, the nopaline-type Ti-plasmids and octopine-type Ti-plasmids are the most predominantly studied. Several studies have also showed that strains exhibit varied host range and some are able to cause tumours in a limited number of species. *Agrobacterium* strain A6 and B6 also shown in Table 4.1 are intimately associated with tumourigenesis in grapevine species. More information about the type of bacterial strains and their plasmid constructs can be found in reports by Knauf et al. (1982), Ma et al. (1987) and Morton and Fugua (2012). These reports and other data show that transformation protocols still need to be updated. This fact has remained so for decades now as it is still poorly understood why some plant species cannot be easily transformed. Already updated transformation models and protocols also differ with the type of *Agrobacterium* strain and plasmid used.

4.2.1 Bacterial Culture

The efficiency of soybean transformation also depends on the nutritional supply of carbon, nitrogen and phosphorus for *Agrobacterium tumefaciens* culture which is

TABLE 4.2
Growth culture media commonly used for rejuvenating *A. tumefaciens*

YEM (1X)	Chemical nutrient (g/L)
Mannitol	5
Yeast extract	0.5
$MgSO_4.7H_2O$	0.2
$NaCl_2$	0.1
$K2HPO_4$	0.5
Sodium gluconate	5
YEP	
Yeast extract	5
Peptone	10
$NaCl_2$	5
Bactor-agar (Solid medium)	12
pH	7.0

rejuvenated before explant infection and co-cultivation (discussed later in this chapter). For this, some commonly used laboratory strains of *A. tumefaciens* (Table 4.1) are reinitiated and regrown using growth culture medium such as Luria-Bertani broth (LB), Yeast Extract Mannitol (YEM) and Yeast Extract Peptone (YEP) medium (Table 4.2). Proper aeration is required for the media to support efficient growth of cells. Unlike during transformation, bacterial isolation from natural sources (soil, plant organs, etc.) critically requires that any media used do not lead to the development of auxotrophic mutants. These are mutants of bacteria unable to synthesise essential compounds or unable to grow in the absence of such particular nutrients in the media. The media compositions upon which the growth of bacteria depends are shown in Table 4.2.

A. tumefaciens strains are being genetically engineered that show auxotrophic properties for plant transformation purposes. The development of a thymidine-dependent growth of *A. tumefaciens* strain LBA4404Thy, EHA101Thy, EHA105Thy and EHA105DThy colonies carrying a *thyA* gene were reported (Aliu et al. 2020). Although not yet available for public laboratories, these strains will be highly desirable to overcome bacterial overgrowth after co-cultivation due to their complete reliance on thymidine as an essential nutrient. The use of YEM and YEP media serves more specific functions such as promoting *vir* gene expressions and surface assembly on infected cells, preferably until genetic introgression takes place (Morton and Fugua 2012).

4.3 ESTABLISHING CONTAMINATION-FREE CULTURES

Among the range of tissue culture-based soybean transformation protocols and procedures used to maintain and grow transgenic plants, aseptic culture establishment remains critically important. Aseptic culture techniques are necessary for the accomplishment of contamination-free bacterial inoculum and infected explant tissues. This stage is used to avoid microbial contamination and provide an *in vitro* environment that promotes stable recovery of transgenic shoots/plantlets without cultures being interfered with by contaminating microorganisms. Such microbial contaminants include bacteria, fungi, moulds, viruses and yeasts. External contaminants are literally present in the air, surfaces of plants, benches, instruments and even on laboratory workers. Furthermore, internal plant structures may also be invaded by systematic virus, virus-like organisms and other internal pathogens. In general, the inclusion of antimicrobial agents such as antibiotics allows for the growth of transformed bacterial cells containing the *goi* over contaminating microbes. Most disarmed *A. tumefaciens* strains used for plant transformation confer resistance to the various antibiotics. This includes the resistance genes located in the bacterial genome, Ti-plasmids and the T-DNA vectors for the broad-spectrum aminoglycoside antibiotics such as gentamicin, kanamycin, rifampicin and streptomycin (Erickson et al. 2014).

4.3.1 Aseptic Seed Cultures

To ensure the success of tissue culture experiments involving soybean transformation, decontamination of plant materials used as explant source and all culture supplies must be attained. Soybean seeds or any other part of the plant to be cultured must be thoroughly sterilised and made free of all contaminating microorganisms before being inoculated on a culture medium. The methods of sterilisation include surface disinfection by physically destroying the microbial contaminants through the application of dry hot air, steam or irradiation (using UV light or gamma irradiations), chemical surface sterilisation with sterilants such as ethanol, mercuric chloride, sodium hypochlorite (domestic bleach) or by filtration (Pierik 1997). The most common sterilisation protocol used for seed sterilisation in soybean transformation cultures is surface decontamination using chlorine gas. This approach may, however, result in excessive dehydration and must be carried out with the minimum 16 hours' duration recommended.

On the other hand, the duration of sterilisation may also depend on the type of plant material used as explant source, regardless of whether they are seeds or not. Sterilisation of both the explant source and explant tissues is mandatory for aseptic culture establishment. Varied responses of germination and seedling health appeared to be adversely affected by prolonged duration of disinfection and the type of sterilant used (Sharma et al. 2018). Chemical sterilants such as hydrogen peroxide (H_2O_2), sodium hypochlorite (NaOCl), ethanol (EtOH) and mercuric chloride ($HgCl_2$) can be very detrimental to explant tissues if their concentrations are not precisely determined. In soybean cultures, very minimal to no contamination incidences can be achieved when seeds are sterilised with NaOCl compared to any other decontaminating agent without any consequential loss of seed viability or seedling vigour. Many studies indicated that any formed culture contamination led to reductions in seed germination, seedling development and health, as well as decreasing the bud or shoot proliferations by explants.

These assertions clearly indicate that the success of a plant tissue culture procedure, whether coupled with plant transformation or not, critically depends on the degree of exclusion of bacteria and other unwanted microorganisms on an *in vitro* culture. Liu et al. (2020) also reported similar findings. In addition, Liu's report highlighted that immersing seeds in a chlorine solution for 2 to 4 hours produced the best sterilisation results for the growth of soybean seed variety Mao-Dou No. 5 to develop tissue culture explants. In some cases, *Agrobacterium* overgrowth can cause contaminations leading to the loss of transformation cultures (Figure 4.1). Unlike the typical microbial growth contamination in tissue culture caused by fungi, other bacteria or any other external or internal microorganisms, this bacterial manifestation is imposed by the bacterium intended for T-DNA delivery into targeted plant tissues.

Agrobacterium overgrowth as it appears in Figure 4.1 is usually aggravated by the exhaustion of culture nutrients and the β-lactam antibiotics aimed at controlling and eliminating the growth of this bacterium. Reports such as those of Zhang et al. (1999), Paz et al. (2006) and Li et al. (2017) demonstrated the use of β-lactam antibiotics such as carbenicillin, cefotaxime, timentin and vancomycin for the suppression and effective growth control of *A. tumefaciens* under *in vitro* culture conditions.

Prior to infection and co-cultivation with *A. tumefaciens* (pTF101.1 vector construct, containing the antibiotic kanamycin and streptomycin-resistant gene), explants were derived from *in vitro* developed seedlings obtained by germinating disinfected soybean seeds on MS basal culture medium supplemented with 3 mg/L benzyladenine (BAP). After co-cultivation, cotyledonary explants were subcultured on MS shoot induction medium containing 2 mg/L BAP, and filter sterilised 50 mg/L cefotaxime as well as 100 mg/L vancomycin. Bud formation and shoot induction were

FIGURE 4.1 *Agrobacterium tumefaciens* overgrowth causing contamination and inhibition of tissue proliferation in double cotyledonary node explants.

observed at least two weeks before culture contamination by *Agrobacterium* can take place. The soybean transformation protocol used was performed as described by Zhang et al. (1999) and Paz et al. (2006) with modifications, which included the type and age of explants as well as the composition of tissue culture media used.

4.3.2 Culture Medium

To efficiently transfer drought stress tolerant, disease and insect resistant genes through genetic transformation in soybean, a good regeneration protocol remains a prerequisite. An effective soybean regeneration system is needed to provide a suitable receptor system for the transfer of *goi* and its expression. Therefore, the optimisation and development of suitable culture media containing the nutrients essential to the plant, as indicated in Table 4.3, are also mandatory in promoting the proliferation of tissues and organs carrying these biotic and abiotic stress-resistant genes. The main goal of using tissue culture medium is to create ideal conditions for obtaining healthy and improved plants in a short period of time and at minimal costs, post-explant infection and co-cultivation with *Agrobacterium tumefaciens*.

Although this is not easy to achieve, many protocols elaborate on the modifications done on culture medium composition (see Table 4.3 and Table 4.4) for efficient regeneration. It is unarguably clear that the development of this tissue culture medium revolutionised research and strengthened many breeding approaches, including transgenic technology. It was the combination of mineral nutrients indicated in Table 4.3

TABLE 4.3
Summarised content (mg/L) of inorganic elements involved in culture medium that are beneficial for the growth of plant tissues and organs *in vitro*

Nutrient	MW (g/mol)	Amount (mg/L)
NH_4NO_3	80.044	1650
KNO_3	101.1032	1900
$CaCl_2.7H_2O$	110.98	440
$MgSO_4.7H_2O$	246.50	370
FeEDTA	344.057	35
KH_2PO_4	136.086	170
H_3BO_3	61.83	6.2
$MnSO_4.H_2O$	169.02	22.3
$ZnSO_4.2H_2O$	161.4	0.25
KI	166.0028	0.83
$Na_2MoO_4.2H_2O$	241.95	0.25
$CuSO_4.5H_2O$	249.69	0.25
$CoSO_4.7H_2O$	281.11	0.03

Source: Pierik 1997.

TABLE 4.4
Carbon source, gelling agents, vitamins and other organic compounds used in plant tissue culture (mg/L)

Organic compounds	MS	Gamborg's B_5
Agar/gelrite	3000	3000
Sucrose	30000	30000
Glycine	2	
Myoinositol	100	100
Nicotinic acid	0.5	5
Thiamine HCl	0.1	10
Pyridoxine	0.5	1
L-cysteine	50–200	50–200
Dithiothreitol	100–200	100–200
Acetosyringone	40–100	40–100
Antibiotics	50–500	50–500
Phytohormones	0.1–5	0.1–5

which showed that the optimal growth and morphogenesis of tissues in different plant species in a culture also vary according to their nutritional requirements. The artificial media used supply the nutrients necessary for plant growth. Inorganic elements like calcium (Ca), magnesium (Mg), nitrogen (N), phosphorus (P) and sulphur (S) are required in relatively large quantities. Small amounts of elements like iron (Fe), nickel (Ni), boron (B) and copper (Cu) are also required (Table 4.3). Therefore, the success of plant tissue culture as a means of regeneration is greatly influenced by the inclusion of both set of nutrients in a culture medium.

4.4 TYPE AND COMPETENCY OF EXPLANTS

Most precursor protocols for *in vitro* soybean transformation use cotyledonary node explants. In a study that produced the first transgenic soybean plants, five (5) day old cotyledons germinated on Gamborg's B_5 medium supplemented with 1.15 mg/L BAP were used. Cotyledons from soybean cultivars Delmar, Maple Presto and Peking were regenerated on B_5BAP medium containing 500 mg/L carbenicillin and 100 mg/L cefotaxime with *Agrobacterium tumefaciens* strain A208 carrying cointegrates of a disarmed nopaline plasmid (pTiT37SE-pMON9749 and pTiT37-SE-pMON894). The plasmids contained a chimeric gene for kanamycin resistance and a gene for glyphosate tolerance (Hinchee et al. 1988). This study and many others tested the competency of cotyledon explant targeted for regeneration and examination of the transformation efficiency using histochemical markers such as the β-glucuronidase (*GUS*) reporter gene system assay to identify transformed cells.

GUS assay identifies transgenic tissues through hydrolytic action of β-glucuronidase enzyme that converts the substrate 5-bromo-4-chloro-3-indolyl glucuronide (X-Gluc) into soluble blue precipitates in the cell cytoplasm. The genes (*uidA* or *GUS*) of this reporter system are incorporated in transformation vectors and get introduced into genetically modified plants to produce valuable indications to identify and track genetic changes in T-DNA recipient cells (Tehryung et al. 1999). Usually, cotyledonary explants or callus associated with the excision wound site prepared for bacterial uptake are widely used. The utilisation of cotyledon explants in targeting T-DNA introgression in soybean plants was then demonstrated more clearly in numerous recent transformation reports involving the use of *A. tumefaciens*. The first improved use of cotyledonary node explants derived from mature seeds instead of immature ones for genetic transformation was reported (Paz et al. 2006). This was followed by the use of half-seed explants for efficient regeneration in soybean transformation (Paz and Wang 2009).

The use of single and double cotyledonary node explants derived from 10-day-old seedling on MS basal culture medium supplemented with 2 to 4 mg/L BAP was then developed (Mangena et al. 2015; Mangena 2021). The development of new alternative explants or the optimisation of those that are already established for an efficient *in vitro* regeneration system remains a prerequisite for soybean transformation. Most of these explants, particularly cotyledonary explants with or without axillary meristems, immature cotyledons and mature cotyledonary nodes, including hypocotyls, are continuously tested for soybean transformation. However, the explants are highly efficient for obtaining higher regeneration frequencies in plant tissue culture. Ample evidence suggests a whole different picture when it comes to *in vitro* soybean transformation. The advantages of obtaining high shoot numbers using these explants for incisions with *A. tumefaciens* are still yet to be realised.

4.5 CO-CULTIVATION OF EXPLANTS WITH *A. TUMEFACIENS*

The purpose of a co-cultivation stage during *Agrobacterium tumefaciens*-mediated genetic transformation is to afford the bacterium sufficient time to infect and exchange genetic materials with targeted host tissues. Far more common, plants always defend themselves against such genetic exchanges. Although plants lack an immune system comparable to animals, they may use chemicals and protein-based defence systems to detect and prevent infections by invading organisms. Similarly, any living plant cell, tissue, organ or part of plant chosen as explant and explant source for *in vitro* genetic transformation could physiologically detect *Agrobacterium* invasion, in addition to sensing tissue injury formed during explant preparation. Since plants include a defence system or hypersensitive reactions would be already triggered and cells bridged through the excision of explants from the explant source, barrier structures such as the cell wall or epidermal cuticle will fail to protect the plant from bacterial invasions.

Furthermore, plant cells will also fail to give disrupted cells the protective strength and rigidity since cell integrity would have been collapsed by the induced mechanical destruction of the cells during explant preparation. Hypersensitive reactions (HR) are a mechanism used by plants to prevent the spread of infections by invading microbial pathogens. HR is commonly characterised by the production of specific secondary metabolites, elicitors and programmed cell suicide (programmed cell death, PCD) or overall growth retardation to induce pathogen resistance (Locato and Gara 2018; Balint-Kurti 2019). Although PCD was first and long identified in plants, this phenomenal mechanism was never investigated in association with induced *A. tumefaciens* plant cell infections. Inducible defence mechanism may also include the production of chemicals toxic to invading organisms or pathogen-degrading enzymes (Freeman and Beattie 2008).

However, *A. tumefaciens* uses specialised genes contained in the Ti-plasmid that are designed and intended to disable these defence systems. In a similar way, the frequency of transformation may be reduced by *Agrobacterium*'s failure to establish intimate connections or suppress plant host defence systems. It has been observed by many researchers that transformation frequencies directly correlate with the duration of co-cultivation of explants with *Agrobacterium*. A significantly high *GUS* expression after co-culturing explants with *A. tumefaciens* for four (4) days was observed, and then the lowest expression was recorded after one (1) day in sugarcane. Composition of the medium used for co-cultivation was reported to have had these significant effects on percentage *GUS* positive callus cells reported (Joyce et al. 2010).

An MS-based embryogenic callus induction medium supplemented with 3 mg/L 2,3-dichlorophenoxy acetic acid (2,4-D) was used in the above study. Similar findings were also made by Paz et al. (2004) and Li et al. (2017). Other factors widely reported to affect co-cultivation and the frequency of transformation include the density of *Agrobacterium* inoculum and inclusion of organic supplements such as acetosyringone as well as dithiothreitol as indicated in Table 4.4. These organic compounds are all included in the Gamborg's B_5/MS-based co-cultivation medium. *A. tumefaciens* inoculum density ranging between 0.8 and 1.0 has always been associated with higher positive transformation rates.

4.6 SHOOT INDUCTION AND ELONGATION

The production of new plant outgrowths in the form of adventitious multiple shoots is the stage that proceeds the co-cultivation of explants with *Agrobacterium tumefaciens*. According to *in vitro* procedures that are normally followed, shoot induction and multiplication can be brought about from the axillary meristems found on the cotyledonary junctions. This is the area on the seedling-derived soybean explants where incisions are made for *Agrobacterium* infections as illustrated previously by Mangena et al. (2017). Prior incision and removal of the epicotyls serve to stimulate the growth of adventitious shoots. Shoot induction depends on this stimulation of precocious axillary bud growth by overcoming the dominance of shoot apical meristems. The production of plants from axillary buds of cotyledons has proved to be the most efficient and generally applicable method. This approach also remains very reliable for the production of true-to-type plantlets in a culture.

Largely from a developmental perspective, transient gene expression associated with meristematic stimulation or shoot formation could also be coupled with T-DNA introgression. The data on cotyledonary explants-based genetic transformation of soybean and other legumes suggest that this direct shoot organogenesis also alleviates somaclonal variations. The formation of shoots in a culture is unique in that both axillary meristems and new shoot meristems can be initiated from differentiated somatic cells than embryogenic cells. However, when adventitious shoots are formed, they may need elongation, followed by adventitious root formation. Elongation involves cell enlargement or expansions naturally regulated by endogenous hormones and environmental stimuli. Some studies, such as those of Fukazawa et al. (2000), reported the expression of a transcriptional activator with a basic leucine zipper (bZIP) domain for the repression of shoot growth (RSG).

The transgenic tobacco plants exhibiting RSG presented severely inhibited stem internode growths and also had less endogenous amounts of gibberellins (GAs). These effects could be restored by the exogenous application of auxins, gibberellins or brassinolides. Unlike cytokinins, these hormones promote the elongation of cells along the longitudinal axis, whereas the expansion of cells along the transverse axis is promoted by cytokinins (Pierik 1997).

4.7 ROOTING OF ELONGATED SHOOTS

The growth and proliferation of adventitious roots in shoot cultures is usually promoted by incorporating plant growth regulation, usually auxins, into the growth medium. Most often such a treatment effectively removes the dominance of apical and axillary shoots controlled by gibberellins and cytokinins and then favours more root formation if possible in large quantities. The root formation is used as miniature cuttings for plant hardening and acclimatisation. In a culture, rooting requires the removal of elongated shoots from the elongation medium and subculture on a fresh rooting medium. The medium may be supplemented with auxins such as naphthalene acetic acid (NAA), indole-3-acetic acid (IAA) or indole-3-butyric acid (IBA). Abscisic acid (ABA) has also been found to stimulate rooting in stem cuttings of mung bean and English ivy. This study showed that ABA partially overcomes the inhibitory effects of gibberellic acid on root formation in mung bean (Chin et al. 1969).

A higher frequency of rooting of 97 and 95% was achieved on soybean shoots regenerated from 1.5 and 2 mg/L BAP, respectively. Rooting took place 7 to 15 days after transfer to MSB_5 medium enriched with 0.1 mg/L IAA or 0.5 mg/L NAA (Soto et al. 2013). The function of this stage is to prepare regenerated shoots for transplanting out of the aseptic, protected environment of *in vitro* cultures to outdoor conditions in the greenhouse or transplanting area. Rooting of elongated shoots could take place under *in vitro* culture conditions or *ex vitro* environment depending on a researcher's preferred protocol.

4.8 ACCLIMATISATION

The adaptation of regenerated plants to soil usually has nothing to do with whether the acclimatised plantlets are not transformed or transgenic. To achieve this stage, all rooted plants must be transferred to the greenhouse or transplanting area under controlled climatic conditions for better plant growth. This step often includes transplanting the plants into plastic pots covered with transparent plastics to create a humidity chamber (Soto et al. 2013). Some commercial or experimental micropropagation systems often avoid *in vitro* rooting, by dipping microcuttings in auxins solution or rooting powder, and inserting them directly into a growth medium for rooting. According to Hartmann et al. (2014) this *ex vitro* procedure does not only provide an excellent transition from the culture environment to open air, but also minimises labour requirements for handling plants under aseptic *in vitro* conditions.

Generally, existing distinctive morphological difference between roots formed *in vitro* compared to *ex vitro* roots may slow down the acclimatisation process in some legume species. In soybean, standardisation of these conditions may influence the final transformation efficiency, especially after rooting as indicated by Zia et al. (2010). Since this stage represents a shift from the heterotrophic to autotrophic mode of plant nutrition, both basal culture nutrients and the carbon source are usually reduced by half. Any standardisation or optimisation is done taking into consideration the sizes of leaves, reduced protective functions, raised and stomata that fail to close, as well as delicate roots. *In vitro* regenerated plants have a reduced leaf cell layer, lack normal amounts of epicuticular waxes on the leaf surface and the stomata fail to close after being removed from the *in vitro* environment (Premkumar et al. 2001; Hartmann et al. 2014).

4.9 SCREENING OF TRANSGENIC PLANTS

Generally, transgenic plant analysis is required to probe the transgene introgression in host tissues using molecular techniques such as polymerase chain reaction (PCR). PCR as well as other methods of analysis are performed to determine the expression of transgenic constructs within the target plant genomes constituting a single transformation event independent of other events. However, challenges in plant transformation often arise when the effects of the transgene cannot be identified among the non-transformed tissues and plants. Furthermore, analysis of transgenic plants also reveals profound effects of DNA methylation taking place during the expression of transgenes or genes of interest. According to Kumar and Fladung (2001), methylations of the DNA, including other intrinsic and extrinsic factors, were found to reduce the stability of transgene expressions, further causing greater difficulties in the screening of transgenic plants.

Transgene repeats, incomplete/complete multiple copies of the transgenes and flanking plant DNA or chromosomal positions also critically influence the transgene inactivation as described by Kumar and Fladung (2001). Unfortunately, this may subsequently lead to either pseudo-positive or negative screening of transformed soybean plants. These challenges also contribute to the problem of recalcitrance to genetic transformation that is observed in soybean and other legume crops, including monocots such as corn and wheat. Moreover, the loss of transgenes creates the need for efficient optimisation and simplification of *in vitro* tissue culture conditions used to detect transgenic plants from non-transformed soybean plants. On this point, the newly established techniques should ensure effective means of achieving long-term stable transgene expression screening which would allow easy detection and analysis of transformed plants under diversely adverse conditions. One of the advantages of using *Agrobacterium*-mediated genetic transformation is that the approach uses relatively less complex and low-cost procedures for probing transgene integrations.

For example, the technique uses antibiotics, especially kanamycin (which is an aminoglycoside), for the preliminary selection and screening of transformants. This class of antibiotics was also found to control and eliminate the growth of *A. tumefaciens* after co-cultivation with explants without adversely affecting shoot proliferation when used at elevated concentrations. This is one of the most critical functions normally performed by β-lactam antibiotics. The β-lactam antibiotics such as carbenicillin, cefotaxime, timentin and vancomycin have been used regularly in *Agrobacterium*-mediated transformation to suppress and eliminate the bacterium after explant infection and co-cultivation. However, culture optimisations are still highly recommended since these antimicrobial agents were also previously associated with serious reductions in the transformation efficiencies. Cefotaxime, for instance, was found to have detrimental effects on the growth of maize callus despite working well to control *Agrobacterium* overgrowth at a concentration of 250 mg/L (Opabode 2006).

In another study, cefotaxime and carbenicillin were both shown to inhibit the regeneration frequency of tomato cotyledonary explants (cvPKM-1) during *Agrobacterium*-mediated transformation even though they showed profound effects on the suppression of the bacterium, particularly at minimal concentrations (Mamidala and Nanna 2009). Finally, an herbicide glufosinate-ammonium is also widely used as a selective agent for the identification of transformed plantlets, either in a culture or by foliar spraying *ex vitro*. A range of 6.0 to 8.0 mg/L glufosinate concentration is normally recommended for the selection of transgenic plants. Its mode of action has been very controversial, but this herbicide irreversibly prohibits the activity of the enzyme glutamine synthetase leading to increased levels of ammonia and deficiency of glutamine in plant tissues. These effects consequently result in the inhibition of photosynthesis and respiration, eventually causing the deaths of plants. Transgenic plants that contain glufosinate ammonium (phosphinothricin) resistant genes (*pat* or *bar* gene) have been produced via *Agrobacterium*-mediated genetic transformation.

4.10 OTHER CONSIDERATIONS

Many studies have been set out to produce transgenic plants exhibiting resistance to various abiotic and biotic stress factors. *Agrobacterium tumefaciens*-mediated genetic transformation has been widely utilised in the efforts to produce stress-resistant plants, especially for agronomically important crops. However, transformation rates remain low, and researchers still endeavour to develop certain additional ways in which the limitations associated with this method could be averted. For instance, the use of aseptic cultures through the sterilisation of explants, culture medium, instruments and use of transfer hoods is mandatory for establishing contamination-free cultures. Aseptic cultures are very important to achieve because the transformation of targeted tissues can be limited and severely affected by contaminating bacteria or fungi, including the *A. tumefaciens* that has been intended for the plant gene expression process as indicated in Figure 4.1. Furthermore, the selection and application of appropriate pre-treatments or preconditioning organic chemicals to supplement basal culture media are also crucial for achieving high transformation efficiencies in soybean. Plant transformation conditions in general, particularly the culturing environment, can be organised, maintained and optimised as discussed below.

4.10.1 Equipment and Laboratory Supplies

The entire process of plant tissue culture-based *Agrobacterium*-mediated genetic transformation remains one of the most unsophisticated laboratory procedures in molecular plant biotechnology. This technique can be carried out in any standard laboratory without any specially designed equipment or functional operating area. Any tissue culture laboratory with sufficient bench area for media and explant source preparation can be effectively used. Appropriate glassware and instrumentation such as culture vessels, Erlenmeyer flasks, test tubes, beakers and a range of pipettes in general are required. Equipment such as pH meters, balance, autoclave, laminar flow hoods, bead sterilisers, centrifuge and a benchtop ultraviolet-visible (UV/Vis) spectrophotometer are necessary. Performing this procedure does not require any specialised laboratory planning or changing the daily running/operations of the laboratory, except that the laboratory should be certified for handling transformed bacteria and transgenic plants.

Although methods like particle bombardment, ultrasound, shockwaves, microprojectiles and electroporation, as well as the use of laser-microbeams are more efficient due to their easy and efficient targeting of plant tissues, the applications of these methods in plant transformation are mainly prohibited by their costs of operations and tedious and more laborious protocols. For diverse groups of scientists and research teams working on plant transformation, some of the several considerations required in developing a genetic transformation laboratory include the following:

1) Low-cost procedures with a large number of transformation events;
2) Technical simplicity with minimum manipulations;
3) Requiring standard facilities to regenerate transgenic plants;
4) Promoting safe operations without transgene escapes or involving dangerous laboratory procedures or substances (Rivera et al. 2012).

4.10.2 Surfactants

The indirect methods of plant transformation, especially genetic manipulations via *Agrobacterium tumefaciens*, involve the use of surfactants. These are compounds used to lower the surface tension and promote interactions between *Agrobacterium* and explant tissues. In simple terms, surfactants help with *A. tumefaciens* attachment to explant tissues and by eliminating unwanted substances on or around the area where excisions are made on the explants. In a report involving soybean, percentages of bud formation and transient expression efficiency were both improved, accounting for enhanced regeneration and transformation efficiency after co-treating cells with sonication and Silwet L-77. Synergistic treatment of Silwet L-77 with sonication improved the transformation efficiency between 2.5 to 5.7% in two soybean genotypes, Jack and Zhonghuang 10 (Bing-fu et al. 2015). Surfactants such as Pluronic acid F68, Silwet L-77 and Tween-20 are widely used in *Agrobacterium*-mediated transformation to enhance T-DNA delivery in targeted tissue explants. However, others like Triton X-100 were found to be highly toxic to plant tissues (Dehestani et al. 2010).

4.10.3 Antinecrotic Treatments

One of the major challenges facing soybean transformation is oxidative burst. This takes place during explant-*Agrobacterium* interactions following infection and co-cultivation stages. Oxidative stress effects are closely associated with contaminations or bacterial overgrowths, explant tissue browning and tissue senescence (Mangena 2021). Oxidative stress results from the by-products of plant aerobic respiration leading to higher expressions of reactive oxygen species (ROS). These species have unpaired electrons that form highly reactive, unstable free radicals that when present in excess become harmful to cells, causing damage to cellular structures, including photosynthetic apparatus. This damage on cellular structures was shown to have adverse effects on the potential of explants to proliferate shoots, induce cell senescence and reduce the developmental rate of tissue predetermined for *de novo* shoot organogenesis for the subsequent production of transgenic plantlets.

The imbalance of ROS and secondary antioxidant metabolites on transformed tissues causes reductions in the transformation efficiencies. Furthermore, ROS have been shown to affect plant cells by altering the plant's physiological and defence response mechanisms (Huang et al. 2017). Moreover, the pre-treatment of cotyledonary node explants prior to or post-infection with *A. tumefaciens* reduces oxidative burst. Some of the most commonly used organic supplements possessing antioxidant potential include ascorbic acid, L-cysteine and dithiothreitol as described in previous sections dealing with the infection and co-cultivation of explants with *A. tumefaciens*. Paz et al. (2006) treated immature soybean explants by including in the culture medium both L-cysteine (400 mg/L) and dithiothreitol (154.2 mg/L) to improve the efficiency of transformation.

Similarly, Enrique-Obregon et al. (1999) earlier tested 15 mg/L ascorbic acid, 40 mg/L cysteine and 2 mg/L silver nitrate for the same intentions in sugarcane transformation. Other studies in soybean, rice, maize and tomato transformation also showed that explant viability and amenability to *A. tumefaciens* were significantly improved when explants were treated with a mixture of these compounds (Bing-fu et al. 2015; Gui et al. 2016; Didone et al. 2018; Ho-Plagaro et al. 2018).

4.11 SUMMARY

Plant transformation through the use of *Agrobacterium tumefaciens* involves a series of the most simplified steps that can be performed in any standard plant tissue culture laboratory. It includes the preparation of plant materials (preconditioning and decontamination of explant and explant source), vector construct preparation, explant infection, co-cultivation, shoot induction and screening as well as selection of transformed plantlets. The genes of interest usually contained within the T-DNA binary vector are integrated and expressed within the host plant cells aided by organic antinecrotic compounds added in the infection and co-cultivation medium. In general, the organic supplements used to promote transformation efficiency include acetosyringone, dithiothreitol, cysteine and PGRs. Acetosyringone was identified as the main additive in transformation cultures, and well known to improve genetic transformation frequencies by up-regulating virulence genes of *A. tumefaciens* (Nakano 2017).

Cotyledon explants are usually infected with *A. tumefaciens* in the presence of a few antioxidant and antimicrobial compounds. These chemical compounds ensure the prevention of oxidative burst, control and eliminate *Agrobacterium* overgrowths in addition to potentially contaminating unknown bacteria. Explant tissues need to be treated with these virulence and growth-promoting chemicals because untreated cells often result in severe tissue senescence. Many reports recorded improved transformation rates after the exposure of explant tissues to acetosyringone, dithiothreitol and cysteine, including surfactants such as Tween-20 and Silwet L-77 (Opabode 2006; Paz et al. 2006; Mamidala and Nanna 2009; Joyce et al. 2010; Zia et al. 2010; Erickson et al. 2014; Bing-fu et al. 2015). In addition to its long-described procedure and advantages as a plant manipulation technology, *A. tumefaciens*-mediated genetic transformation can also be coupled with techniques such as sonication to effect genetic changes between the bacterium and plant explants as described by Amal et al. (2020).

4.12 ABBREVIATIONS

ABA	Abscisic acid
B_5	Gamborg's B5 medium
BAP	Benzyladenine
DNA	Deoxyribonucleic acid
EtOH	Ethanol
GAs	Gibberellins
GUS	β-glucuronidase
H_2O_2	Hydrogen peroxide
HgCl	Mercuric chloride
HR	Hypersensitive reaction
IAA	Indole-3-acetic acid
LB	Luria-Bertani broth
MS	Murashige and Skoog medium
NAA	1-Napthaleneacetic acid
NaCl	Sodium chlorite
NaOCl	Sodium hypochlorite
PCD	Programmed cell death

PGRs	Plant growth regulators
PCR	Polymerase chain reaction
ROS	Reactive oxygen species
RSG	Repression of shoot growth
T-DNA	Transferred deoxyribonucleic acid
Ti-plasmid	Tumour-inducing plasmid
UV	Ultraviolet
Vis	Visible light
YEM	Yeast extract mannitol
YEP	Yeast extract peptone

REFERENCES

Aliu E, Azanu MK, Wang K and Lee K. (2020). Generation of thymine auxotrophic *Agrobacterium tumefaciens* strains for plant transformation. *BioRxiv* (Reprint publication), 1–20.

Amal TC, Karthika P, Dhandapani G, Selvakumar S and Vasanth K. (2020). A simple and efficient *Agrobacterium*-mediated in planta transformation protocol for horse gram (*Macrotyloma uniflorum* Lam. Verdc.). *Journal of Genetic Engineering and Biotechnology* 18(9), 1–9.

Balint-Kurti P. (2019). The plant hypersensitive response: Concepts, control and consequences. *Molecular Plant Pathology* 20(8), 1168–1178.

Bing-fu G, Yong G, Jun W, Li-Juan Z, Long-Guo J, Hui-Long H, Ru-Zheng C and Li-Juan Q. (2015). Co-treatment with surfactant and sonication significantly improves *Agrobacterium*-mediated resistant bud formation and transient expression efficiency in soybean. *Journal of Integrative Agriculture* 14(7), 1242–1250.

Chin T-Y, Meyer MM and Beevers L. (1969). Abscisic acid-stimulated rooting of stem cuttings. *Planta* 88, 192–196.

Dehestani A, Ahmadian G, Salmanian AH, Jelodar NB and Kazemitabar K. (2010). Transformation efficiency enhancement of *Arabidopsis* vacuum infiltration by surfactant application and apical inflorescence removal. *Trakia Journal of Sciences* 8(1), 19–26.

Didoné AD, de Silva RM, Ceccon CC, Teixeira T, Suzin M and Grando MF. (2018). Increased transient genetic transformation in immature embryos of Brazilian BR 451 maize co-cultivated with *Agrobacterium tumefaciens. Acta Scientiarum* 10, e36475. 1–6.

Enriquez-Obregon GA, Prieto-Samsonov DL, de la Riva GA, Perez MI, Selman-Housein G and Vazquz-Padron RI. (1999). *Agrobacterium*-mediated Japonica rice transformation a procedure assisted by an antinecrotic treatment. *Plant Cell Tissue and Organ Culture* 59, 159–168.

Erickson JL, Ziegler J, Guevara D, Abel S, Klosgen RB, Mathur J, Rothstein SJ and Schattat NH. (2014). *Agrobacterium*-derived cytokinin influences plastid morphology and starch accumulation in *Nicotiana benthamiana* during transient assays. *BMC Plant Biology* 14(127), 1–20.

Freeman BC and Beattie GA. (2008). An overview of plant defenses against pathogens and herbivores. *The Plant Health Instructor* 94, 1–14.

Fukazawa J, Sakai T, Ishida S, Yamaguchi I, Kamiya Y and Takahashi Y. (2000). Repression of shoot growth, a bZIP transcriptional activator regulates cell elongation by controlling the level of gibberellins. *Plant Cell* 12(6), 901–915.

Gui H, Li X, Liu Y and Li X. (2016). Evaluation of factors impacting *Agrobacterium*-mediated indica rice transformation of IR58025B- a public maintainer line. *Journal of Rice Research* 4(163), 1–7.

Hartmann HT, Kester DE, Davies FT and Geneve RL. (2014). *Hartmann and Kester's Plant Propagation: Principles and Practices*, 8th eds. Person Education Limited, Edinburgh Gate. pp. 683–684.

Hinchee MAM, Connor-Ward DV, Newell CA, McConnell RE, Sato SJ, Gasser CS, Fischhoff DA, Re DB, Fraley RT and Horsch RB. (1988). Production of transgenic soybean plants using *Agrobacterium*-mediated DNA transfer. *Biotechnology* 6, 915–922.

Ho-Plágaro T, Huertas R, Tamayo-Navarrete MI, Ocampo JA and GarcíaGarrido JM. (2018). An improved method for *Agrobacterium rhizogenes*-mediated transformation of tomato suitable for the study of arbuscular mycorrhizal symbiosis. *Plant Methods* 14(34), 1–12.

Horn J and Fries W. (2018). Detection of collapse and crystallization of saccharide, protein and mannitol formulations by optical fibres in lyophilisation. *Frontiers in Chemistry* 6(4), 1–9.

Huang H, Ullah F, Zhou DX, Yi M and Zhao Y. (2017). Mechanisms of ROS regulation of plant development and stress responses. *Frontiers in Plant Science* 10(800), 1–10.

Joyce P, Kuwahata M, Turner N and Lakshmanan P. (2010). Selection system and co-cultivation medium are important determinants of *Agrobacterium*-mediated transformation of sugarcane. *Plant Cell Reports* 29(2), 172–183.

Knauf VC, Panagopoulos CG and Nester EW. (1982). Genetic factors controlling the host range of *Agrobacterium tumefaciens*. *Phytopathology* 72, 1545–1549.

Knauf VC, Panagopoulos CG and Nester EW. (1983). Comparison of Ti-plasmids from 3 different biotypes of *Agrobacterium tumefaciens* isolated form grapevines. *Journal of Bacteriology* 153(3), 1535–1542.

Kumar S and Fladung M. (2001). Gene stability in transgenic aspen (*Populus*). II. Molecular characterisation of variable expression of transgene in will and hybrid aspen. *Planta* 213, 731–740.

Li S, Cong Y, Liu Y, Wang T, Shuai Q, Chen N, Gai J and Li Y. (2017). Optimisation of *Agrobacterium*-mediated transformation in soybean. *Frontiers in Plant Science* 8(246), 1–15.

Liu Y, Li X, Liu K, Rao G and Xue Y. (2020). Optimisation of aseptic germination system of seeds in soybean (*Glycine max* L.). *Journal of Physics Conference Series* 1637, 1–5.

Locato V and De Gara L. (2018). Programmed cell death in plants: An overview. In De Gara L and Locato V (eds), *Plant Programmed Cell Death. Methods in Molecular Biology*, Vol 1743. Humana Press, New York. pp. 1–8.

Ma D, Yanofsky MF, Gordon MP and Nester EW. (1987). Characterization of *Agrobacterium tumefaciens* strains isolated from grapevine tumors in China. *Applied and Environmental Microbiology* 53(6), 1338–1343.

Mamidala P and Nanna RS. (2009). Influence of antibiotics on regeneration efficiency in tomato. *Plant Omics Journal* 2(4), 135–140.

Mangena P. (2021). Effect of *Agrobacterium* co-cultivation stage on explant response for subsequent genetic transformation in soybean (*Glycine max* (L.) Merr.). *Plant Science Today* 8(4), 905–911.

Mangena P, Mokwala PW and Nikolova RV. (2015). *In vitro* multiple shoot induction in soybean. *International Journal of Agriculture and Biology* 17, 838–842.

Mangena P, Mokwala PW and Nikolova RV. (2017). Challenges of *in vitro* and *in vivo* *Agrobacterium*-mediated genetic transformation in soybean. In Kasai M (eds), *Soybean: The Basis of Yield, Biomass and Productivity*. IntechOpen, Rijeka, Croatia. pp. 75–94.

Morton ER and Fugua C. (2012). Unit 3D.1 laboratory maintenance of *Agrobacterium*. *Current Protocol in Microbiology* 24(3D.1.1-3D.1.16), 1–8.

Nakano Y. (2017). Effects of acetosyringone on *Agrobacterium*-mediated transformation of *Eustoma grandiflorum* leaf disks. *AARQ* 51(4), 351–355.

Opabode JT. (2006). *Agrobacterium*-mediated transformation of plants: Emerging factors that influence efficiency. *Biotechnology and Molecular Biology Review* 1(1), 12–20.

Paz M and Wang K. (2009). Soybean transformation and regeneration using half-seed explants. US Patent No. 7, 473, 822 B, 1-7. January 6, 2009.

Paz MM, Martinez JC, Kalvig AB, Fonger TM and Wang K. (2006). Improved cotyledonary-node method using an alternative explant derived from mature seed for efficient *Agrobacterium*-mediated soybean transformation. *Plant Cell Reports* 25, 206–213.

Paz MM, Shou H, Guo Z, Zhang Z, Banerjee AK and Wang K. (2004). Assessment of conditions affecting *Agrobacterium*-mediated soybean transformation using the cotyledonary node explant. *Euphytica* 136, 167–179.

Pierik RLM. (1997). *In Vitro Culture of Higher Plants*, Vol 45. Martinus Mishoff Publishers, Dordrecht, UK. pp. 89–100.

Prakash O, Nimonkar Y and Shouche YS. (2013). Practice and prospects of microbial preservation. *FEMS Microbiology Letters* 339(1), 1–9.

Premkumar A, Mercado JA and Quesada MA. (2001). Effects of *in vitro* tissue culture conditions and acclimatisation on the contents of rubisco, leaf soluble proteins, photosynthetic pigments and C/N ratios. *Journal of Plant Physiology* 158, 835–840.

Rivera AL, Gomez-Lim M, Fernandez F and Loske AM. (2012). Physical methods for plant transformation. *Physics of Life Reviews* 9, 308–345.

Sharma S, Kaur M, Goyal R and Gill BS. (2014). Physical characteristics and nutritional composition of some new soybean (*Glycine max* (L.) Merrill) genotypes. *Journal of Food Science and Technology* 51(2), 551–557.

Sharma V, Kumari S, Sharma S and Jadon VS. (2018). Seed sterilisation in soybean for establishment of aseptic cultures. *Journal of Emerging Technologies and Innovative Research* 5(12), 958–964.

Soto N, Ferreira A, Delgado C and Enriquez GA. (2013). *In vitro* regeneration of soybean plant of the Cuban incasoy-36 variety. *Biotechnologia Applicada* 30(1), 34–38.

Tehryung KIM, Chowdhury MKU and Wetzstein HY. (1999). A quantitative and histological comparison of GUS expression with different promoter constructs used in microprojectile bombardment of peanut leaf tissue. *In Vitro Cellular and Developmental Biology-Plant* 35, 51–56.

van Asma FM. (1995). Growth and storage of *Agrobacterium*. In Gartland KMA and Davey MR (eds), *Agrobacterium Protocols: Methods in Molecular Biology*. Springer, Totowa. pp. 1–7.

Zhang Z, Xing A, Staswick P and Clemente T. (1999). The use of glufosinate as selective agent in *Agrobacterium*-mediated transformation of soybean. *Plant Cell Organ Culture* 56, 37–46.

Zia M, Rizmi ZF, Rehman RU and Chaudhary MF. (2010). Agrobacterium mediated transformation of soybean (*Glycine max* L.): Some condition standardization. *Pakistan Journal of Botany* 42(4), 2269–2279.

5 *In Vitro* Cultures Commonly Used for Plant Transformation

5.1 INTRODUCTION

As discussed in Chapter 4 plant tissue culture (PTC) plays an important role in the advancement of research for basic and applied plant sciences, breeding, the production of natural or pharmaceutical products and the propagation and recovery of transgenic plants. This technology has proven to be immensely successful for the micropropagation of many horticultural and ornamental plants, reducing the amount of time taken by breeders to introduce new cultivars. PTC has been used to extend a range of plant genotypes that can be propagated *in vitro* (Hartmann et al. 2014), and to efficiently achieve the propagation of threatened, endangered and recalcitrant plant species. Among the objectives that this technology attempts to achieve is the propagation of genetically manipulated plants. The application in genetic engineering is intended to produce true-to-type genetically engineered clones of plants regenerated from cells, tissues or organs containing genes of interest.

Plants derived from plant tissue culture are called microplants and can be generated using techniques such as axillary shoot formation, seedling culture, callus culture and somatic embryo formation. Axillary shoot formation involves methods like meristem culture, nodal culture and shoot cultures. Axillary branching or what is also termed lateral shoot formation is formed from the junction between the stem and cotyledonary leaves. Recently, a plant regeneration protocol was optimised for soybean cultivars NARC-4 and NARC-7 using cotyledonary nodes as explants to induce axillary shoots. The frequency of shoot regeneration ranged between 54 and 88% on Gamborg's B5 medium containing varying concentrations (0–2 mg/L) of 6-benzylamino purine, zeatin riboside and kinetin (Zia et al. 2016). Seedling cultures, however, include embryo rescue used for breeding interspecific crosses that had normally failed to set viable seeds, and immature ovules or ovaries that get fertilised with pollen *in vitro*.

Haploid cultures such as ovule and ovary culture, including pollen cultures, involve the differentiation of a gametophyte stage to sporophytic development used for the production of hybrid cultivars, often without sexual recombination (Takeshita et al. 1980). PTC techniques depend on the plant cell's *totipotency* driving the development of tissues to form shoot buds or primordial buds which subsequently grow into intact plants. The theorisation of the 'cell theory' in relation to *totipotency* postulated by Matthias Schleiden and Theodor Schwann in 1838 was comprehensively reviewed by Trigiano and Gray (2005), and recently alluded to by Hartmann et al. (2014).

DOI: 10.1201/b22829-5

In most cases, callus, shoot and meristem cultures use plant cells' *totipotency* in the production of transgenic plants through *Agrobacterium tumefaciens*-mediated genetic transformation.

A simple method for obtaining transgenic soybean callus tissues was reported by Hong et al. (2007). Infection efficiency of *A. tumefaciens* and higher regeneration frequencies were achieved using organogenic callus derived from cotyledonary node explants possessing axillary buds as the starting materials. However, meristem and shoot cultures of uninterrupted and organised cell growths remain the most preferred method for the recovery of transgenic plants compared to callus cultures.

These cultures give rise to small, organised shoots that can be directly elongated and rooted *in vitro*. In addition, the use of meristematic tissues (either small apical or lateral apices) also ensures the establishment of plants that are free of virus infections. Both the use of small and larger stem apices or lateral buds in meristem and shoot culture play a significant role in plant propagation. Therefore, *in vitro* cultures are continuously considered the most efficient and adaptable procedures for efficient *Agrobacterium*-mediated gene transfers in soybean. This chapter, thus, reviews and summarises the role of callus, shoot, protoplast and meristem culture for genetic manipulation in soybean using *A. tumefaciens*. This chapter comprehensively reviews these cultures by focussing the discussions on the purpose of culture, requirements and applications for the efficient recovery of transgenic plants.

5.2 CALLUS CULTURE

A mass of cells derived from any part of the plant material of either meristematic or non-meristematic tissue origin that can be cultured under *in vitro* conditions is referred to as callus. This mass of dedifferentiated cells can be caused to divide on medium supplemented with relatively high amounts of auxins or efficiently on a combination of auxin and cytokinins. In plants, callus cells are induced in response to the wounding of tissues. The establishment of callus cells using any explant implies that the tissue serves as a competent source for obtaining an undifferentiated mass of cells. In plant tissue culture and other different *in vitro*-based applications, callus cultures are established and used for various purposes as discussed further in this chapter. Callus cultures are generally used for research, breeding and genetic improvement studies. These forms of culture use any vegetative tissues as explant source and are referred to as stationary cultures.

Callus formation can also be used to obtain suspension and protoplast cultures (Patel and Patel 2013). The latter is established by isolating single cells with enzyme digested cell walls, and the other involves the agitation of multiplying or dividing cells in a hormone-supplemented liquid medium. Frequently, the suspension cultures are used in elucidating the growth and developmental processes of single plant cells.

5.2.1 PURPOSE OF CALLUS CULTURE

The establishment of callus, suspension and single cell cultures using the undifferentiated mass of cells is used for a number of functions as indicated in Figure 5.1.

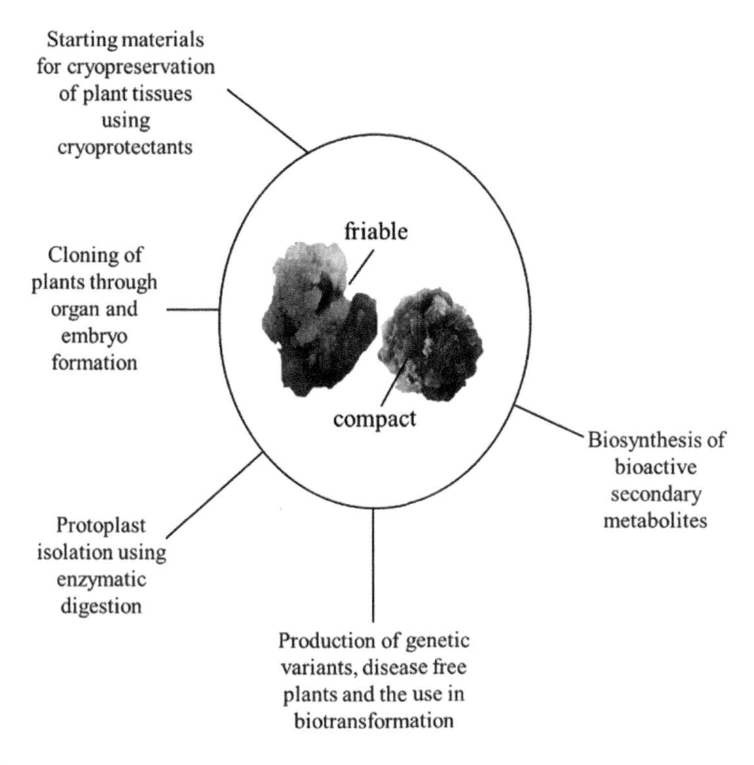

FIGURE 5.1 The basic type of common callus cell forms/structures used for potential application in different tissue culture systems.

The recent development of genetic engineering of plants to produce plants with desirable features has indeed added enormous amounts of new and growing benefits. Likewise, callus cultures offer a wide range of beneficial uses in agriculture, pharmacology and pharmacy, especially by the synthesis of bioactive secondary metabolites. This approach, combined with genetic engineering and specialised PTC techniques, is used to produce chemical drugs and therapeutic recombinant proteins, including antibiotics (Efferth 2019). The secondary metabolites produced through callus and suspension cultures comprise neutral compounds for the manufacturing of fragrances, food additives, flavourants, colourants and agricultural chemicals. But nutraceuticals, pharmaceuticals and cosmetic products remain preferably derived from cell cultures as raw materials (Wang et al. 2017).

Cryogenic cooling is crucial for the storage of shoot apices and somatic embryos or cells. Cryopreservation as it is known may be applied through freeze-induced dehydration of cells and methods based on vitrification, involving the treatment of plant cell samples with cryoprotectants. According to Bekheet et al. (2020), vitrification steps include the treatment of plant tissues with a highly concentrated solution of cryoprotectants, rapid cooling, sample rewarming, followed by the removal of cryoprotectant solution and the recovery of plant samples. This method is broadly applied in callus cultures, suspension cultures and many other somatic plant cultures. On the other hand, protoplasts isolated from callus tissues are usually obtained by digestive enzymes. Cellulases and hemicellulases digest the cellulosic component of the plant cell walls, together with pectinase enzymes that function to break down the pectins that cement and hold adjacent cells together. Cellulases, hemicellulases and pectinases used for plant cell digestion are extracted from various sources, such as fungi, snail and termite gut, and are available commercially.

Furthermore, callus-based protoplast culture has been used to generate variants (somaclones) with improved characteristics, bypassing any intraspecific and interspecific incompatibility barriers that may limit the process. As a first step in some transformation experiments, callus induction from the primary explant can be genetically manipulated or proliferated from genetically modified explants for shoot regeneration. Many reports allude to the usefulness of these cultures in counteracting many constraining factors involved in PTC techniques and genetic engineering (Opabode 2017).

5.2.2 Requirements for Establishing Callus Culture

The three criteria in callus culture establishment are similar to those found in many other *in vitro* cultures such as (1) the aseptic preparation of plant materials, (2) the selection of suitable nutrient medium and (3) the incubation of culture under controlled environmental conditions. The initiation of callus culture may use small pieces of leaf, root and stem tissues which can be subcultured on appropriate culture medium to rapidly produce these undifferentiated cells. However, it is well known that the nutrient culture medium used also favours the rigid growth of bacteria and fungi. These microbes can contaminate the culture, inducing toxins that severely affect the cultured tissues and exhausting the medium. This observation justifies the requirement of sterilisation or disinfecting technique during callus culture establishment. The sterilisation of plant material may involve both internal and external treatments in the attempt to reduce all types of contaminations.

As indicated by Mihaljevic et al. (2013) surface sterilisation is the most important step in the preparation of explants for any tissue culture purposes controlling bacterial and fungal contaminations. This is one of the many reports strongly emphasising the importance of completely removing the exogenous and endogenous contaminating microbes with sodium hypochlorite, calcium hypochlorite, ethanol, mercuric chloride, hydrogen peroxide and silver nitrate being the most commonly used disinfectants. Efficient callus induction (78.3–88.8%) using three varieties of soybean from shoot tip and cotyledonary node explants on MS medium containing 3.0 mg/l 2,4-Dichlorophenoxyacetic acid (2,4-D) in combination with 1.0 mg/l 6-benzylaminopurine (BAP) was reported (Islam et al. 2017). The explants were sourced from *in vitro* germinated seeds of soybean cultivar BARI soybean-5 (BS-5), BARI soybean-6 (BS-6) and Shohag that were surface sterilised using 70% (v/v) ethanol and 40% clorex with 1–2 drops of Tween-20 and savlon.

Joyner et al. (2010) involved the use of 2,4-D and naphthalene acetic acid (NAA) separately at a concentration of 0–21 μM, and in combination for callus induction in soybean. Abundant amounts of callus cells were obtained according to the study, either from the cotyledon in the medium modified with 2,4-D or 2,4-D combined with NAA. Not all cells on the explant contributed to callus formation, but the level of plant growth regulators (auxins, cytokinins, ethylene, gibberellins, etc.) plays a major role in the induction of calli cells.

5.2.3 Application in Plant Transformation

Callus cultures are among the most important tissue culture techniques in basic applied science and commercial applications, in addition to their use in genetic engineering. In all major crop species used for commercial agriculture, this method has been used in various ways to contribute to the growth and yield characteristics of all varieties of useful plants. These include economic crop plants such as wheat, rice, maize, cotton, soybean and sugarcane, etc. A high frequency of callus induction at 2 mg/l 2,4-D, varied between 63.3 and 92.37 was reported in three Indian varieties, namely Sita, Masuri and Rapali (Upadhyaya et al. 2015). In wheat and maize, callus cells were successfully induced from MS cultures using mature and immature embryos as explants, respectively (Turhan and Baser 2004; Malini et al. 2018). Unlike many other studies, Khamrit et al. (2012) reported the use of yeast extract and coconut water in addition to the 2,4-D and NAA plant hormones for callus initiation and maintenance in sugarcane.

The use of explants similar to those reported above was also reported by Rajeswari et al. (2010) for callus formation in maize. All these reports, including Islam et al. (2017) and Joyner et al. (2010) that demonstrated callus induction in soybean, emphasised the role of callus initiation in *in vitro* plant regeneration under sterile conditions. Often, this method can be coupled with plant transformation and suspension culture rather than being used in micropropagation technique alone as reported by Malini et al. (2018) (Figure 5.2). Callus cells are also responsible for establishing competent cells that are amenable to stable gene transfer using *Agrobacterium*-mediated gene technology or any other suitable alternative for generating new variants (Figure 5.2).

These callus cultures have been frequently used for species presenting genetic barriers and recalcitrance for plant transformation during *in vitro* regeneration. Several reports have already established that the responses of most cultures are genetically controlled. Moreover, the ability to maintain callus lines for several years, especially through vitrification, made possible the ability to retain cell differentiation. Although these methods face different specific challenges, such as the difficulty of removing protective tissues and low frequency of recovery in protoplast or suspension cultures, more procedures are being developed to isolate and culture genetically enhanced plant cell lines.

5.3 SHOOT CULTURE

The main objective of plant tissue culture systems is to successfully produce a large number of healthy plants in a short period of time, and at minimal expense. Micropropagation through tissue culture remains one of the most interesting and widely used approaches in plant biotechnology. The various applications as indicated

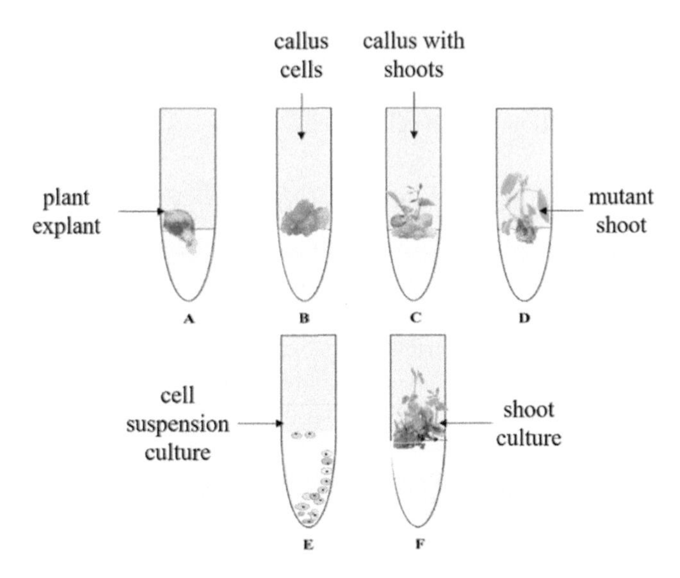

FIGURE 5.2 Demonstration of soybean callus culture. Single cotyledonary node used for establishing callus (A), infected explant showing induced friable calli cells (B), shoot initiation (C), recovered mutant shoots from *Agrobacterium*-infected explants (D), established suspension cell culture (E) and shoot culture from callus cells (F).

in Figure 5.2 (F) can also be performed without any intervening callus stage during micropropagation. Shoot cultures are the most preferred form of tissue culture that uses different explant types to regenerate (propagate) new plants, especially when coupled with plant genetic transformation. As previously indicated, the establishment of this technique formed the basis for many plant improvement protocols, and the elimination of undesirable characteristics accompanying conventional breeding methods of several agronomically important crops. *In vitro* regeneration protocols now serve as key objects of genetic improvement and are also used to initiate cultures suitable to facilitate other varieties of *in vitro* cultures, including callus and protoplast culture.

5.3.1 Purpose of Shoot Culture

By far, shoot cultures are the most frequently used micropropagation systems across laboratories worldwide. These cultures have a simple purpose of maintaining plant tissues in a stabilised state and multiply the microshoots to the number required for elongation, rooting and subsequent plant regeneration stages (Hartman et al. 2014). This has been widely used for the rapid production of genetically uniform, pathogen-free and high-quality plants, both on a small and commercial scale. Therefore, shoot cultures are mostly preferred for supplying the production demand for most micropropagated plant species (Victorio et al. 2012). As indicated by the above report, this approach has led to increased standardisation of many protocols aimed at propagating large quantities of plant species used for pharmaceutical and other processing industries. These also include the enhancement of quantities of materials used for the preservation of genetic resources, such as cryopreservation or *ex situ* conservation and in the breeding of newly improved varieties.

For instance, shoot cultures have been successfully used to propagate plants that are difficult to grow vegetatively, as it naturally occurs in orchids. The propagation of orchids via plant tissue culture is necessary to benefit from their medicinal and ornamental value. A lot of propagation techniques were developed for *Phalaenopsis* spp., *Dendrobium* spp., *Miltassia* sp. and their hybrids through shoot culture from explants such as leaf segments, flower stalk, flower stalk nodes, callus and seeds (Utami and Hariyanto 2019; Datta et al. 2018; Oliveira et al. 2019). Other applications necessary for the establishment and use of shoot cultures include axillary branching as a tool to clone plants, the production of shoot buds for use in cryopreservation to create gene banks and the mass production of any important species through efficient *in vitro* regeneration protocols.

5.3.2 Culture Requirements for Efficient Shoot Multiplication

A complete plant regeneration system through shoot culture also relies on the interactions of a number of factors/steps/stages in plant tissue culture. Like any other cultures, disinfection of contaminants from the surface of the organ used as explants remains of paramount importance. The explant surfaces must be completely free of contaminants because they are inoculated on a tissue culture medium containing nutrients and sugar that support microbial growth as mentioned previously (Hartmann et al. 2014). Furthermore, this kind of culture requires fully balanced and appropriately adjusted hormone levels/combinations to successfully develop a complete regeneration system for any particular species. There is also a strong reliance on the interaction between cytokinins and auxins. Although cytokinins alone are used for shoot culture establishment, without a combination with auxins, the latter are used in very low concentrations, especially during the *in vitro* regeneration of woody and recalcitrant species.

As is already fundamentally known, cytokinins stimulate shoot development and growth, while auxins favour root initiation and growth. Additionally, the use of competent tissue explants, especially from young and highly dividing cells, remains a prerequisite for an efficient *in vitro* regeneration of soybean and other plant species. Meristem and shoot tip culture consisting of an undifferentiated group of actively dividing cells capable of multiplying throughout their life is used for this purpose. The basic growth of these cells is supported by PGRs which direct the developmental responses of tissues in culture, promoted by minimum levels of cytokinins, while shoot elongation is inhibited by the use of higher amounts of these cytokinins.

However, hormone requirements may vary at different stages of culture such that variable growth responses may occur during consecutive subcultures. The adjustment of cytokinin concentrations may be necessary where explant response is relatively low. Moreover, the use of seeds as explant source requires that high seed viability, including seedling vigour and seedling age from which explants are derived, be carefully considered, to avoid relying upon cytokinin/auxin adjustments. So, both the quality of explant and culture medium composition strongly influence *in vitro* shoot culture responses.

5.3.3 Application of Shoot Culture in Plant Transformation

Intensive commercial micropropagation of clones or mutant varieties requires proper optimisation of not only the medium but also other culture elements, which can be done systematically by testing a wide range of concentrations individually or combined. Such elements include inorganic nutrients, pH, growth regulators, explant quality, genotype and culture maintenance/incubation conditions. Current research on crop improvement still emphasises culture standardisation of these abovementioned factors as the most predominant efficiency-controlling elements widely affecting *in vitro* regeneration for subsequent genetic improvement. However, the growing interest in crop improvement has focussed attention on the benefits associated with targeted crops and the use of tissue culture-based technology for enhancing the desired crop traits (Brown and Thorpe 1995; Sedeek et al. 2019).

As reported by many researchers, recalcitrance to genetic transformation in soybean is said to be due to factors such as the low infection rates, poor regeneration capacity and genotype specificity. Low infection rates are caused by poor quality of explants (often dependent on the age and nature of the explant), the regeneration efficiency depends on the culture conditions, and lastly, genotype specificity is inherent to the type of plant used, together with the level of reproducibility of the protocol in varieties within the same species or genus. Shoot proliferation is considered more simple and straightforward under *in vivo* culture conditions than *in vitro*. Nonetheless, the use of cotyledonary node explants developed from soybean seedlings germinated on a medium supplemented with PGRs (preferably BAP) could facilitate high competency of multiple bud and shoot induction for subsequent plant regeneration and genetic manipulations.

This is the main reason why achieving plant regeneration via cotyledonary node system remains widely practiced in *in vitro* plant tissue culture systems of many legumes and recalcitrant seeds, including other cereal species of agronomic importance. As indicated previously by Mangena et al. (2017), the use of this approach with the aid of solid media containing cytokinins may enable potential embryogenic tissues found on the cotyledonary junctions or any meristematic region to take up and express the transgenes via *Agrobacterium*-mediated genetic transformation. It should, however, be noted that transgenic soybean shoots have been successfully produced through this transformation technique using both immature and mature cotyledonary node explants bearing undifferentiated axillary meristematic cells at the coty-node junctions (Olhoft and Somers 2001; Patel et al. 2013).

5.4 MERISTEM CULTURE

All plants contain the meristems that generate cells responsible for primary and secondary growth. These sets of perpetually undifferentiated cells give rise to plant

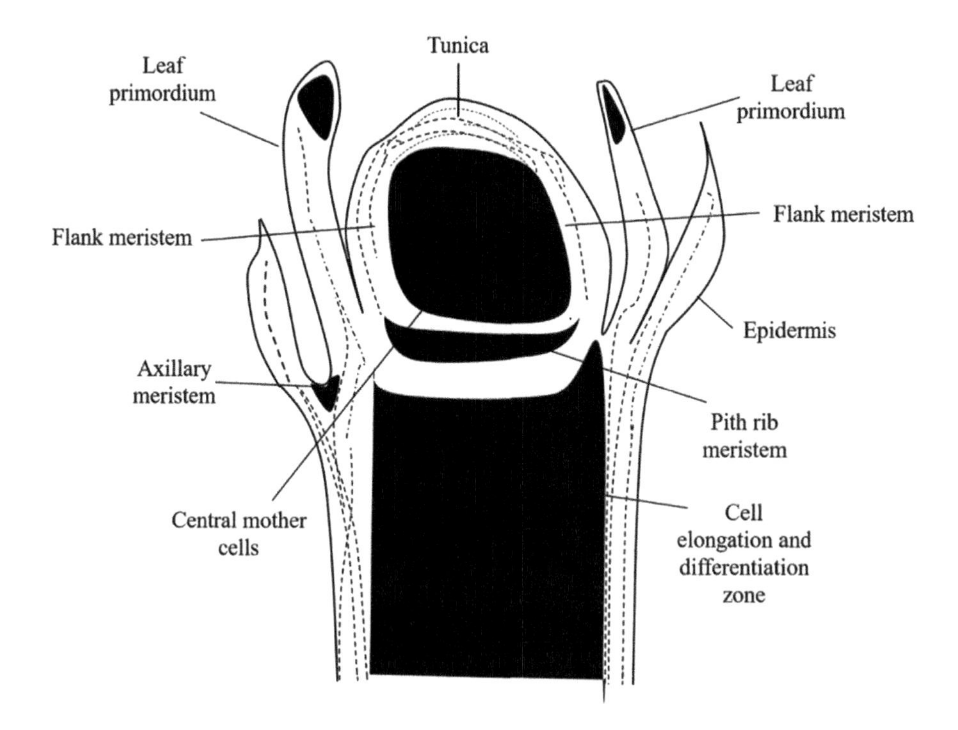

FIGURE 5.3 The terminal bud illustration showing apical meristem, with early developing leaves and different layers of meristem.

cells and tissues that develop into mature organs (Figure 5.3). Unlike animal cells, these cells ensure that plant growth is not limited to the embryonic or juvenile period, but instead, growth occurs throughout the plant's life cycle. At any given time, the flank, peripheral and central cell lines divide into cells when conditions permit, leading to new cells that can differentiate into new tissue systems or organs (Figure 5.3). Therefore, researchers dealing with meristem cultures target such actively dividing cells under plant tissue culture conditions. This type of culture involves the removal of meristematic tissues from a central dome located inside the apical and lateral buds and placing them on the culture media for the development of the whole plant.

The explants used in this type of culture constitute the meristematic domes which are sometimes subcultured with small subtending leaf primordia. Hartmann et al. (2014) highlighted that the number of additional vegetative structures that are included in the culture also depends on the length of the excised meristematic tissue.

Plant propagation using this approach has been used for a number of plant species to produce clean virus-free plants. Figure 5.4 depicts culture procedures of soybean that feature the use of shoot tips, and could be considered as a typical eudicot terminal bud culture to produce improved plants recovered through *in vitro* plant regeneration. Often, this type of culture involves an indirect and intervening callus stage for successful plant propagation. This technique has been used mostly in herbaceous underground and above-ground nodal or axillary shoot propagation in numerous

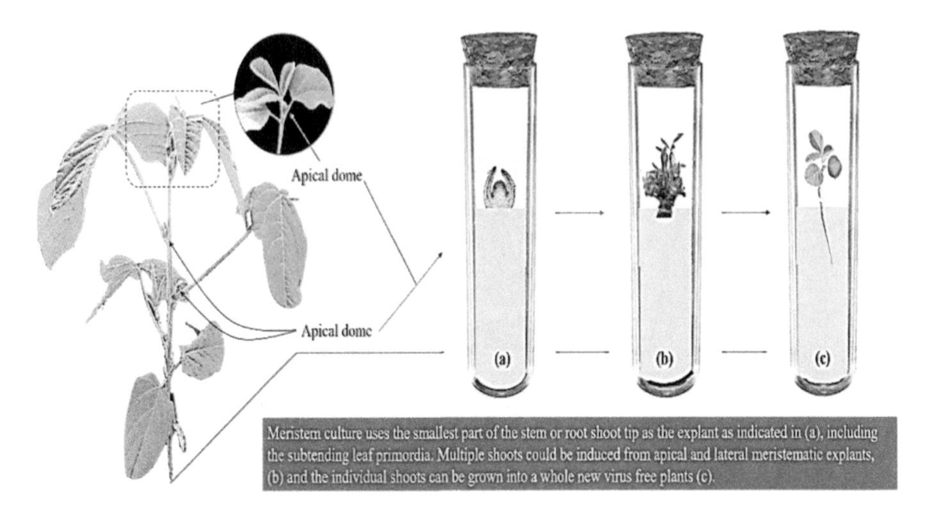

FIGURE 5.4 Examples of different steps in the isolation and subculture of apical/lateral meristematic tissue explants for *in vitro* propagation in soybean.

crop and ornamental plants. The technique has proved to be more complex and less efficient with woody plants, except when using lateral and terminal meristems that are usually exploited in axillary shoot culture. Axillary shoot culture serves as another type of shoot culture involving stool shoots, nodal and axillary branching in *in vitro* plant propagation cultures. Amongst such cultures that utilise lateral and apical meristems are the proliferation of pseudocorms, minitubers and micrografting as reported by Tantasawat et al. (2015), Gami et al. (2013) and Badalamenti et al. (2016) in *Dendrobium*, *Solanum* and *Palecyphora*, respectively.

5.4.1 Purpose of Meristem Culture

The establishment of meristem cultures involves the use of small bud explants that hold undifferentiated actively dividing cells, without differentiated provascular and vascular tissues. This approach is preferred because it provides inherent genetic stability and may be generally used in the production of bacteria, fungi and virus-free plants. Apart from pathogen elimination (particularly viruses), this approach was used in mass propagation, germplasm storage and genetic transformation of numerous monocots and dicots species of economic importance. Successful applications of shoot tip and meristem cultures were reported in fodder grasses and cereals, including the development of the industrial production of secondary metabolites using cell suspension culture obtained from these explants. *In vitro* regeneration for the micropropagation of *Indica* rice variety Jaya was achieved on MS basal medium using multiple shoot buds. In another study, the multiplication of shoot primordia generated and maintained from mature embryos of hybrid rice (*Oryza sativa* L.) was also reported (Sandhu et al. 1995; Yoshida 1996).

Other studies include a robust protocol for the efficient rapid propagation of *Sorghum bicolor* (L.) Moench hybrids (NC+262, NC+6C21 and NC+6B50) reported by Sadia et al. (2010). Meristem-tips of two Iranian *Ficus carica* L. (cultivar Jaami-ekan and Sabz) were used for shoot proliferation, root induction and subsequent plant regeneration (Sahraroo et al. 2019). In legumes (soybean, cowpea, peanut, chick-pea and common bean), the regeneration potential of shoot apical meristems was also reported using agar-solidified MS nutrient medium supplemented with various concentration of benzyladenine (BA) and naphthaleneacetic acid (NAA) alone or in combination (Kartha et al., 1981). Furthermore, the growth and development of initiated buds were also reported by Benedicic et al. (1997) in *Phaseolus vulgaris* L. cv. Zorin using shoot and callus culture formation.

The production of disease-free plants was achieved by separating shoot buds from the internal disease-infected tissues. This as well as other reports have indicated that several bacterial, fungal and viral diseases are transmitted through the vascular and other tissue systems. Since regeneration from meristems results in plantlets that grow to maturing, despite the size of explant used, this approach could be methodically exploited for the transformation of legume crops.

5.4.2 Meristem Culture Requirements

As indicated previously, micropropagation by meristem cultures is also used commercially because it allows the rapid propagation of plants in large quantities. Meristem cultures can be efficiently and rapidly established using tissues isolated from stem/root tips. However, a slow growth and pace of regeneration are normally observed due to the size of explants used. Explants that are very small in size usually demonstrate poor growth response and survival rate. These difficulties in shoot proliferation necessitate the improvement of culture procedures, especially for subsequent *Agrobacterium*-mediated genetic transformation. Furthermore, meristem cultures follow stages similar to those involved in micropropagation through callus and shoot cultures. They include culture initiation, shoot induction/multiplication, the rooting of the developed shoots and the hardening of the regenerated plantlets.

The culture initiation step primarily requires explant transfer to a medium for callus induction before the formation of new shoots. This implies that selected explants must be subcultured on callus induction medium after the necessary preconditioning and surface disinfection. An organised apex of the root/shoot (typically less than 1 mm in length) from the selected donor is excised and subcultured for *in vitro* shoot induction. The culture conditions are regulated to allow for the organised growth of the apical meristem tissues directly into shoots. Conditions must be controlled in such a way that no intervention of any adventitious organs takes place because some of the meristematic cells are genetically predetermined to form certain plant organs (Grout 1990).

5.4.3 Use of Meristem Culture for Plant Improvement

Most importantly, meristem cultures are very beneficial in the elimination of plant viruses and other related microbial pathogens during the production of a large number of *in vitro* propagated plants, but are rarely looked at for plant transformation. So far, meristem culture serves as the main method used for plant virus elimination purposes both in vegetative propagation and under *in vitro* conditions. As previously discussed, shoot and root apical meristem comprise the apical dome where cells are free of pathogens due to the lack of connectivity to other tissue systems that may accumulate those pathogens. In a comparative study of meristem culture technique with shoot tips for obtaining virus-free garlic plants using different media compositions and RT-PCR (real-time polymerase chain reaction) for virus detection (Taskin et al. 2013), it was reported that the garlic plants propagated via this culture did not show any viruses despite the fact that over 70% viral particles were detected on the plants.

Pests and diseases are common in *Allium* species as a result of contaminated vegetative materials used for propagation because of the sexual sterility. Meristem cultures are, therefore, used to obtain virus-free plants than transgenic plants because there are no effective control methods, especially directly against viruses and other pathogens like nematodes. Shoot cultures are the most often used micropropagation systems, especially when coupled with plant transformation rather than meristem cultures which were initially developed for a similar purpose as well (Hartmann et al. 2014). The comparison between meristem cultures and techniques such as axillary shoot, adventitious shoot cultures, etc., indicates clearly that meristems are more suitable for application in disease-free programmes instead of trying to achieve more successes in micropropagation. The fact that vegetatively propagated plants can systematically infect their entire population with pathogens, easily passing from one generation to another, together with transformation barriers, explains why more focus has been placed mainly on disease elimination than genetic improvement.

Furthermore, latent viruses with no detectable symptoms may severely affect the entire population of clonal variety with the disease. In soybean, various researchers have rather exploited embryogenic or meristematic cell axes for the efficient regeneration of transgenic plants (Dang and Wei 2007; Rech et al. 2008; Guo et al. 2015; Sato et al. 2016). However, it is worth noting that despite these efforts, the development of transgenic soybeans is still not a routine process due to its recalcitrance and the various other transformation barriers, including the genotype-linked quality of explants and poor cell proliferation post-infection with *Agrobacterium tumefaciens*.

5.5 PROTOPLAST CULTURE

Callus culture serves as an important tissue culture system that can be subcultured and maintained indefinitely to provide cells used to establish protoplast culture. Protoplast culture involves the removal of the cell wall surrounding the protoplasm which is a living component of the plant cell. Cells comprising of exposed protoplasmic contents hold the potential to obtain somatic hybrids to create genetic variability. This technique is also used in transformation studies to obtain transgenic plants.

5.5.1 Purpose of Protoplast Culture

A number of applications involving the use of protoplast culture include the creation of somatic hybrids, genetic transformation and chromosome fragment fusion to create the desired genetic variants. This method is primarily coupled with callus culture establishment (Figure 5.5), particularly when implementing it as a tool for plant genetic improvement. Chabane et al. (2007, 2010) reported the isolation of protoplasts from embryogenic calli in 'Deglet nour' and 'Takerboucht' genotypes. More than 1.5×10^6 protoplast cells were recovered for the micropropagation of *Phoenix dactylifera* L. using calli cells from these genotypes. The use of somatic hybridisation through the fusion of protoplasts enables the genomic combination of different genetic characteristics from different species. This technique has been reported as one of the most powerful tools for the induction of genetic variability in plants. According to Bajaj (1994) somatic hybridisation is used to achieve gene flow in the synthesis of useful growth and reproductive traits from sexually incompatible species.

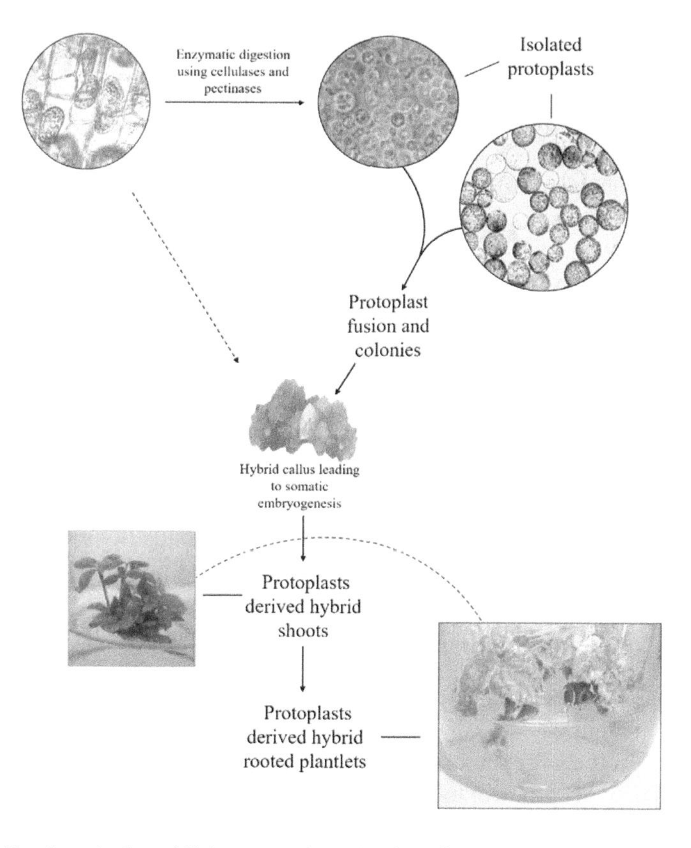

FIGURE 5.5 A typical establishment and application of protoplast culture for use in hybrid shoot and plantlet establishment using callus culture.

The attainment and use of somatic hybrids or cybrids of agricultural importance to achieve better crop growth and yields have also been reported. This technique has been very important in *Citrus* breeding, including various other crops of agronomic value such as tomato, potato, lettuce, rice, cabbage, alfalfa and millet. Fruit trees include species of *Prunus*, *Pyrus*, *Populus* and *Citrus* as already indicated (Ohgawara et al. 1991; Bajaj 1994; Shuro 2018).

5.5.2 Requirements for Protoplast Culture

A study by Ohgawara et al. (1991) reported somatic hybridisation from nucellar cell suspension protoplast culture. This study used navel orange with mesophyll protoplasts of *Troyer citrange* cultured on hormone-free Murashige and Tucker medium containing 0.6 M sucrose. This study, as well as many others on the recovery of somatic hybrid plants using protoplast cultures after Carlson et al.'s report, indicated some of the most essential modifications on this technique since 1972. The study by Carlson et al. (1972) used *Nicotiana glauca* and *N. langsdorffi* leaf mesophylls for protoplast isolation using 4% cellulase solution, 0.4% macerozyme and 0.6M sucrose at pH 5.7. Protoplast density of more than 5×10^7 per mL was reported for both genotypes. According to the above report, all previous studies failed to regenerate calli cells using protoplast isolated from these two species (as exemplified in Figure 5.5) until the stated vigorous calli growth of hybrid genetic combinations obtained on hormone-free medium.

However, culture conditions have been significantly improved, resulting in a number of protocols being developed that entail a callus intervening stage in early 1972. Even though protoplast fusion is currently achieved widely across species, and between varieties of distant and close relatives, each technique/protocol still requires proper optimisation on its culture conditions. Factors that need to be taken into serious consideration are the tissue type being used for protoplast isolation, plant genotype, culture conditions and the growth conditions of the source plants. These factors may serve as serious technical barriers to the development of protoplast cultures, especially for the efficient recovery of genetically improved somatic hybrids. Decades ago, protoplast cultures were also successfully developed in soybeans using the enzymatic treatment of seedling hypocotyls and cotyledonary tissues (Lin et al. 1987; Hammatt and Davey 1988).

Polyethylene glycol-electroporation technique and osmolality reduction regimes for liquid cultures were used to produce protoplast density of over 5×10^4 per mL in both the above-mentioned studies, respectively. Wu and Hanzawa (2018) reported the production of large quantities of high-quality protoplasts using young unifoliate soybean leaves. In another study, the fusion of mesophyll protoplast cells to form heterokaryons of rapeseed and soybean was reported (Kartha et al. 1974). The protoplast system requires efficient culture optimisations in order to provide a versatile and effective protocol for plant improvement. This is particularly required to enable it to serve as a model protocol for the examination of complex regulatory and signalling mechanisms involved during growth and development, especially for anatomical, physiology and biochemical responses of soybeans exposed to biotic and abiotic stress conditions.

5.5.3 Applications in Plant Transformation

Protoplast culture has been widely used for plant transformation in various laboratories worldwide. An earlier study by Lin et al. (1987) reported DNA transfer into protoplasts using polyethylene glycol-electroporation by means of callus culture. This report also relied upon digested competent calli cells to achieve protoplast transformation. Other researchers found protoplast culture preferable for plant transformation because cell walls may serve as barriers blocking the passage of foreign DNA incorporation into the targeted host tissues. Genetic transformation strategies involving the use of protoplasts as a foreign DNA reservoir include electroporation, microinjection and chemically induced (polyethylene glycol-based) gene transfer methods (Kartha et al. 1981). A wide range of species have been reported for successful protoplast isolation and subsequent genetic transformation. In soybean, protoplast isolation and application for transient gene expression has been analysed.

Polyethylene glycol-calcium mediated transformation was optimised to achieve high transformation frequencies using unifoliate protoplasts. A GFP-fusion protein expression was achieved by the transformation of soybean cv. Williams 82 protoplasts using a small plasmid vector p2GWI7-E1. *Agrobacterium tumefaciens*-based binary vectors are usually avoided for protoplast transformation due to their larger sizes of vectors and gene copy number, which may impede high transformation efficiencies (Wu and Hanzawa 2018). Xiong et al. (2019) also reported the transient expression of cytoplasmic GMCRY1 photobody-like proteins using soybean mesophyll protoplasts transformed with a cloned cDNA of GmCRYs into the pA7-YFP vector to generate pA7-GmCRY-YFP plasmid. However, efficient, simple and rapid protoplast isolation protocols have been designed to recover totipotent protoplast cells from soybean tissues.

Limited literature currently exists dating between 2000 and 2021 on the use of protoplast cultures for plant genetic transformation. Barriers contributing to the current decreases in the application of protoplast cultures on genetic engineering may include the longer culture maintenance and establishment periods, compared to the use of direct organogenic cultures, as well as the high costs of operations and requirement of highly skilled personnel to prepare protoplasmic tissues for the development and propagation of transgenic plants (Page et al. 2019).

5.6 SUMMARY

Plant tissue culture forms part of the advanced biotechnological approaches capable of propagating plants using both sexual and asexual means. Among the many existing PTC applications, including those that are covered in this chapter, are techniques such as direct and indirect somatic embryogenesis. According to scant literature, there are indications that plants have used asexual embryogenesis or somatic embryogenesis during the course of evolution to overcome and improve their fitness against environmental stress (Salaun et al. 2021). Somatic embryogenesis shares various developmental and physiological similarities with zygotic embryogenesis. This similarity in growth characteristics led to the wide range use of embryogenic cultures in clonal propagation, regeneration and genetic transformation of many agronomically important crops. Apart from this, here the chapter provided an overview of the main *in vitro* culture methods that have been widely tested for *Agrobacterium tumefaciens*-mediated genetic transformation in soybean.

Meristem, shoot and callus derived protoplasts continue to be tested for the *in vitro* manipulation of soybean despite their many disappointing outcomes in transformation rates. Findings made in many studies still strongly signal the need for a newly established or optimised tissue culture protocol for use in the genetic improvement of soybean under *in vitro* conditions. However, these techniques remain widely used due to the high proliferative capacity of explants, the simplicity of the procedure and the higher regeneration frequency demonstrated before the involvement of *A. tumefaciens* in culture. Furthermore, all mechanisms (molecular, anatomic, morphological and physiological) controlling plant regeneration in relation to *Agrobacterium*-mediated genetic transformation are well understood. It is now common knowledge that high regeneration frequency is a prerequisite for efficient and improved rates of genetic transformation. Axillary shoot formation is, therefore, a preferred culture method for use in *Agrobacterium*-based transformation.

The main objective of these cultures should be to improve transformation efficiency using the minimum amount of time possible, produce a larger number of healthy transgenic shoots and carry out genetic improvement at low costs. Shoot regeneration is achieved without intervening callus. Although callus culture is also frequently used in plant transformation, this procedure is very sensitive to the type and amount of PGRs or other additives used, and it is prone to undesirable culture variations. This results in culture problems such as the production of translucent, succulent plant tissue appearance (known as vitrification or hyperhydricity) in callus suspension cultures, habituation and chimeric effects. These may cause deviations from the original genome makeup, posing serious challenges to *in vitro* cultures because these changes are permanent and may lead to the detection of pseudo-positive screening of transgenic plants. Under normal circumstances, with or without transformation the problem may not appear or be detected until after plant hardening and transplanting (Hartmann et al. 2014).

Nevertheless, plant tissue culture provides researchers with the opportunity to manipulate the culturing environment to suit their study objectives and the needs of the species to be cultured. True-to-cultivar or true-to-type microplants that are free from internal as well as external pathogens can also be established. Finally, PTC allows the optimisation of *in vitro* culture conditions to further enable deviations from the original protocols in order to save costs, promote rapid regeneration and increase the efficiency of transformation while adhering to the general standards of micropropagation.

5.7 ABBREVIATIONS

BAP 6-benzylamino purine or benzyladenine
BA Benzyladenine or 6-benzylamino purine
cDNA Complementary deoxyribonucleic acid

cv.	Cultivar
2,4-D	2,4-Dichlorophenoxyacetic acid
DNA	Deoxyribonucleic acid
GFP	Green fluorescent protein
M	Molarity
mL	Millilitre
MS	Murashige and Skoog medium
NAA	1-Naphthaleneacetic acid
PGRs	Plant growth regulators
PTC	Plant tissue culture
RT-PCR	Real time polymerase chain reaction

REFERENCES

Badalamenti O, Carra A, Oddo E, Carmi F and Sajeva M. (2016). Is *in vitro* micropropagation a valid alternative to traditional micropropagation in Cactaceae? *Palecyphora aselliformis* as a case study. *Springer Plus* 5(201), 1–4.

Bajaj YPS. (1994). Somatic hybridisation: A rich source of genetic variability. In Bajaj YPS (ed), *Somatic Hybridization in Crop Improvement 1. Biotechnology in Agriculture and Forestry*, vol 27. Springer, Berlin, Heidelberg, pp. 3–32.

Bekheet SA, Sota V, El-Shabrawi HM and El-Minisy AM. (2020). Cryopreservation of shoot apices and callus cultures of globe artichoke using vitrification method. *Journal of Genetic Engineering and Biotechnology* 18(2), 1–8.

Benedicic D, Ravnikar M and Gogala N. (1997). The regeneration of bean plants from meristem culture. *Phyton* 37(1), 151–160.

Brown DC and Thorpe TA. (1995). Crop improvement through tissue culture. *World Journal of Microbiology and Biotechnology* 11(4), 409–415.

Carlson PS, Smith HH and Dearing RD. (1972). Parasexual interspecific plant hybridization. *Proceedings of the National Academy of Science* 69(8), 2292–2294.

Chabane D, Assani A, Bouguedourn N, Haicour R and Ducreux G. (2007). Induction of callus formation from difficile date palm protoplasts by means of nurse culture. *Comptes Rendus Biologies* 330(5), 392–401.

Chabane D, Bouguedourn N and Assani A. (2010). Importance of protoplast culture in the genetic improvement of date palm (*Phoenix dactylifera* L.). *Acta Horticulturae* 882, 185–192.

Dang W and Wei Z. (2007). An optimised *Agrobacterium*-mediated transformation for soybean for expression of binary insect-resistance genes. *Plant Science* 173(4), 381–381.

Datta A, Zahara M, Boonkorkaev P and Mishra A. (2018). Effects of plant growth regulators on the growth and direct shoot formation from leaf explants of the hybrid *Phalaenopsis* 'pink'. *Acta Agriculture Slovenica* 111(1), 5–6.

Efferth T. (2019). Biotechnology applications of plant callus cultures. *Engineering* 5(1), 50–59.

Gami RA, Parmar SK, Patel PT, Tank CJ, Chauhan RM, Bhadauria HS and Solanki SD. (2013). Microtuberization, minitubers formation and *in vitro* shoot regeneration from bud sprout of potato (*Solanum tuberosum* L.) cultivar k.badshah. *African Journal of Biotechnology* 12(37), 5640–5647.

Grout BW. (1990). Meristem-tip culture. *Methods in Molecular Biology* 6, 81–91.

Guo B, Guo Y, Hong H, Lin L, Zhang L, Change R-Z, Lu W, Lin M and Qiu L-J. (2015). Co-expression of G2-EPSPS and glyphosate acetyl transferase GAT gene conferring high tolerance to glyphosate in soybean. *Frontiers in Plant Science* 6(847), 1–11.

Hammatt N and Davey MR. (1988). Isolation and culture of soybean protoplasts. *In Vitro Cellular and Developmental Biology* 24, 601–604.

Hartmann HT, Kester DE, Davies FT and Geneve RL. (2014). *Hartmann and Kester's Plant Propagation: Principles and Practices*, 8th eds. Person Education Limited, Edinburgh Gate. pp. 683–684.

Hong HP, Zhang H, Olhoft P, Hill S, Wiley H, Toren E, Hillebrand H, Jones T and Cheng M. (2007). Organogenic callus as the target for plant regeneration and transformation via Agrobacterium in soybean (*Glycine max* (L.) Merr.). *In Vitro Cellular and Developmental Biology- Plant* 43, 558–568.

Islam N, Islam T, Hossain MM, Bhattacharjee B, Hossain MM and Islam S. (2017). Embryogenic callus induction and efficient plant regeneration in three varieties of soybean (*Glycine max*). *Plant Tissue Culture and Biotechnology* 27(1), 41–50.

Joyner EY, Boykin L-SS and Lodhi MA. (2010). Callus induction and organogenesis in soybean [*Glycine max* (L.) Merr.] cv. Pyramid from mature cotyledons and embryos. *The Open Plant Science Journal* 4, 18–21.

Kartha KK, Gamborg OL, Constabel F and Kao KN. (1974). Fusion of rapeseed and soybean protoplasts and subsequent division of heterokaryocytes. *Canadian Journal of Botany* 53(11), 2435–2436.

Kartha KK, Pahl K, Leung NL and Mroginski LA. (1981). Plant regeneration from meristems of grain legumes: Soybean, cowpea, peanut, chickpea and bean. *Canadian Journal of Botany* 59(9), 1671–1679.

Khamrit R, Jaisil P and Bunnag S. (2012). Callus induction, regeneration and transformation of sugarcane (*Saccharum officinarum* L.) with chitinase gene using particle bombardment. *African Journal of Biotechnology* 11(24), 6612–6618.

Lin W, Odell JT and Schreiner RM. (1987). Soybean protoplast culture and direct gene uptake and expression by cultured soybean protoplasts. *Plant Physiology* 84(3), 856–861.

Malini S, Anandakumar CR, Gnanam R and Ramakrishnan HS. (2018). Effect of hormone on callus induction in maize (*Zea mays* L.). *Journal of Applied and Natural Science* 10(1), 202–209.

Mangena P, Mokwala PW and Nikolova RV. (2017). Challenge of *in vitro* and *in vivo Agrobacterium*-mediated genetic transformation in soybean. In Kasai M (ed), *Soybean: The Basis of Yield, Biomass and Productivity*. Intech Open, Croatia, pp. 75–94.

Mihaljevic I, Dugalic K, Tomas V, Viljevac M, Pranjic A, Cmelik Z, Puskar B and Jurkovic Z. (2013). *In vitro* sterilisation procedures for micropropagation of oblacinska sour cherry. *Journal of Agricultural Sciences* 58(2), 117–126.

Ohgawara T, Kibayashi S, Ishii S, Yoshinaga K and Oiyama I. (1991). Fertile fruit tress obtained by somatic hybridization: Navel orange (*Citrus sinensis*) and Troyer citrange (*C cinensis* x *Poncirus trifoliata*). *Theoretical and Applied Genetics* 81, 141–143.

Olhoft PM and Somers DA. (2001). L-cysteine increase *Agrobacterium*-mediated T-DNA delivery into soybean cotyledonary-node cells. *Plant Cell Reports* 20, 706–711.

Oliveira BC, Oliveira MEBS and Cardoso JC. (2019). Feasibility of the new method for orchid *in vitro* rooting using liquid and chemical sterilised culture medium under different sucrose concentration. *Ornamental Horticulture* 25(3), 263–269.

Opabode JT. (2017). Sustainable mass production, improvement and conservation of African indigenous vegetables: The role of plant, tissue culture, a review. *International Journal of Vegetable Science* 23(5), 438–455.

Page M, Pavy MAJ and Carmo-Silva E. (2019). A high-throughput transient expression system for rice. *Plant Cell and Environment* 42(7), 1–8.

Patel M, Dewey RE and Qu R. (2013). Enhancing *Agrobacterium tumefaciens* efficiency of perennial ryegrass and rice using heat and high maltose treatment during bacterial infection. *Plant Cell, Tissue and Organ Culture* 114, 19–29.

Patel MB and Patel RS. (2013). Effects of plant growth regulators (PGRs) on callus induction from leaf segments of Tecomella undulata (Sm.) seem: An important medicinal plant. *International Journal of Scientific and Research Publications* 3(12), 1–4.

Rajeswari S, Muthuramu S, Chandirakala R, Thiruvengadam V and Raveendran TS. (2010). Callus induction, somatic embryogenesis and plant regeneration in cotton (*Gossypium hirsutum* L.). *Electronic Journal of Plant Breeding* 2(4), 1186–1190.

Rech E, Vianna G and Aragao F. (2008). High efficiency transformation by biolistics of soybean, common bean and cotton transgenic plants. *Nature Protocol* 3(3), 410–418.

Sadia B, Josekutty PC, Potlakayala SD, Patel P, Goldman S and Rudrabhatla SV. (2010). An efficient protocol for culturing meristems of Sorghum hybrids. *International Journal of Experimental Botany* 17, 177–181.

Sahraroo H, Zarei A and Babalar M. (2019). *In vitro* regeneration of the isolated shoot apical meristem of two commercial fig cultivars 'Sabz' and 'Jaami-e-Kan'. *Biocatalysis and Agricultural Biotechnology* 17, 743–749.

Salaun C, Lepiniec L and Dubreucq B. (2021). Genetic and molecular control of somatic embryogenesis. *Plants* 10(7), 1467, 1–16.

Sandhu JS, Gosal SS, Gill MS and Dhaliwal HS. (1995). Micropropagation of *Indica* rice through proliferation of axillary shoots. *Euphytica* 81, 139–142.

Sato N, Delgado C, Hernandez Y, Rosabal Y, Ferreira A, Pujol M, Aragao FJL and Enriquez GA. (2016). Efficient particle bombardment-mediated transformation of Cuban soybean (INCA Soy-36) using glyphosate as a selective agent. *Plant, Cell, Tissue and Organ Culture* 128, 187–196.

Sedeek KEM, Mehas A and Mahfouz M. (2019). Plant genome engineering for targeted improvement of crop traits. *Frontiers in Plant Science* 10(114), 1–16.

Shuro AR. (2018). Review paper on the role of somatic hybridization in crop improvement. *International Journal of Research Studies in Agricultural Sciences* 4(9), 1–8.

Takeshita M, Kato M and Tokumasu S. (1980). Application of ovule culture to the production of intergeneric or interspecific hybrids in *Brassica* and *Raphanus*. *Japanese Journal of Genetics* 55(5), 373–387.

Tantasawat PA, Khairum A, Arsakit K, Poolsawat O, Pornbungkerd P and Kativat C. (2015). Effects of different culture media on growth and proliferation of Dendrobium 'Earsakul' protocorm-like bodies. *HortTechnology* 25(5), 681–686.

Taskin H, Baktemur G, Kurul M and Buyukalaca S. (2013). Use of tissue culture technique for producing virus-free plant in garlic and their identification through real-time PCR. *The Scientific World Journal* 781282, 1–5.

Trigiano RN and Gray DJ. (2005). *Plant Development and Biotechnology*. CRC Press, Boca Raton. pp. 301–308.

Turhan H and Baser I. (2004). Callus induction from mature embryo of winter (*Triticum aestivum* L.). *Asian Journal of Plant Sciences* 3, 17–19.

Upadhyaya G, Sen M and Roy A. (2015). *In vitro* callus induction and plant regeneration of rice (*Oryza sativa* L.) var 'Sita', 'Rupali' and 'Swarna Masuri'. *Asian Journal of Plant Science and Research* 5(5), 24–27.

Utami ESW and Hariyanto S. (2019). *In vitro* seed germination and seedling development of a rare Indonesian native orchid *Phalaenopsis amboinensis* J.J.Sm. *Scientifica* 8105138, 1–7.

Victorio CP, Lage C-LS and Sato A. (2012). Tissue culture techniques in the proliferation of shoots and roots of *Calendula officinalis*. *Revista Ciencia Agronomica* 43(3), 539–545.

Wang J, Li J-L, Li J, Liu S, Huang L and Gao W. (2017). Production of active compounds in medicinal plants: From plant tissue culture to biosynthesis. *Chinese Herbal Medicines* 9(2), 115–125.

Wu F and Hanzawa Y. (2018). A simple method for isolation of soybean protoplasts and application to transient gene expression analyses. *Journal of Visualised Experiments* 131(e57258), 1–7.

Xiong L, Li C, Li H, Lyn X, Zhao T, Liu J, Zuo Z and Liu B. (2019). A transient expression system in soybean mesophyll protoplasts reveals the formation of cytoplasmic GmCRY1 photobody-like structures. *Science Chana Life Sciences* 62(8), 1070–1077.

Yoshida T. (1996). *In vitro* propagation of hybrid rice (*Oryza sativa* L.) 1. 'Tissue-cultured' shoot primordial. *JARQ* 30, 1–8.

Zia M, Rizvi ZF, Rehman R and Chaudhary MF. (2016). Short communication. Micropropagation of two Pakistani soybean (*Glycine max* L.) cultivars from cotyledonary nodes. *Spanish Journal of Agricultural Research* 8(2), 448–452.

6 Current and Longstanding Challenges Facing Soybean Transformation

6.1 INTRODUCTION

Soybean, like many other legumes, contains newly improved varieties that are being tested and slowly introduced into the farming system in order to increase crop productivity under adverse environmental conditions. A limited number of cultivars established through genetic engineering have successfully made it to the global market, thus providing farming communities with ample opportunities to immensely benefit from bioengineered varieties that are developed through a precise and controlled addition of new genes (Liu 1997). Such genetically modified plants have new traits that include fungal resistance, nematode resistance, herbicide tolerance and tolerance to drought stress as well as other abiotic stress factors. However, genetic engineering of soybean still occurs at very low rates, and if successfully achieved in one genotype, then it later proves to be genotype specific in another. Genotype specificity means that a protocol used for an efficient transformation of one variety cannot be equally applied across all soybean genotypes.

This phenomenon is referred to as the 'recalcitrance' to genetic transformation (Mangena 2019). Although significant strides have been made in the past decade for specific genotype improvement, recalcitrance remains a major obstacle for growth manipulation of soybeans via *Agrobacterium tumefaciens*. This chapter will, therefore, discuss some of the most pertinent current and longstanding challenges that negatively influence the rate of genetic transformation in this crop. Furthermore, information is very limited to explain why genetic transformation in soybean is still very inefficient, genotype specific and inconsistent among all continuously tested genotypes. This crop remains difficult to transform regardless of the newly improved *in vitro* culture protocols, types of explants and various additives tested and previously proved to achieve significant frequencies of genetic improvement in a few *Glycine* genotypes, other legumes and cereal crop species.

DOI: 10.1201/b22829-6

6.2 LONGSTANDING FACTORS INFLUENCING SOYBEAN TRANSFORMATION

The longstanding factors affecting soybean transformation conducted via indirect *Agrobacterium tumefaciens*-based gene transfer include the genotype, explant type and co-cultivation medium. Reductions in transformation frequencies can be displayed at many levels, but factors indicated above constitute some of the most serious effects observed in the genetic manipulation of plants. However, many studies propose that the optimisation of culture conditions can simplify plant transformation and improve its rates (Paz et al. 2006; Zhang et al. 2014; Verma et al. 2014). Culture standardisations have been commonly used to minimise the effects of the aforementioned factors by, for instance, using mature/immature cotyledonary node explants and respectively, increasing the co-cultivation and infection periods to 5 days and 1 hour. These factors are said to frequently complicate genetic transformation, further delineating the contributions made by the *vir* genes and other promoters of T-DNA expression in the host cells. It was, therefore, envisaged that the optimisation of factors mentioned above during genetic transformation could subsequently account for the lower regeneration efficiencies reported, and improve the recovery of transgenic plants to at least over 70% in soybeans and other legume crops.

6.2.1 GENOTYPE COMPETENCY AND REGENERABILITY

Pre-transformation activities such as genotype evaluation and optimisation of culture conditions could be key steps to starting with *in vitro* regeneration and recovery of transgenic plants. Phenotype assessment and any other basic information about the plant will be beneficial for protocol modification purposes. This indicates why the characteristics of soybean germplasm (involving superior, elite or inferior genotypes) have been crucial for the development of new elite cultivars as they may serve as a reservoir for achieving more genetic diversity. Zhang et al. (2014) and Paz et al. (2006) both identified soybean cultivar William as the most amenable and transformable genotype through *Agrobacterium tumefaciens*-mediated genetic transformation. In this regard, the identification of numerous genotypes showing high affinity and amenability for *A. tumefaciens* could significantly improve the transformation of soybean extensively

According to Jacob et al. (2016), *Glycine* and *Phaseolus* serve as some of the genera containing more accessions of about 229,944 and 261,963 in the world's germplasm banks, respectively. This implies that the collections can serve a key function of providing basic information for breeding programmes. The International Centre for Tropical Agriculture (CIAT), Institute of Crop Germplasm Resources – Chinese Academy of Agricultural Science (ICGR-CAAS), International Crops Research Institute for the Semi-Arid Tropics (ICRISAT), Australian Medicago Genetic Resources Centre (AMGRC), Australian Temperate Field Crops Collection (ATFCC), International Centre for Agricultural Research in the Dry Areas (ICARDA) and International Institute for Tropical Agriculture (IITA) are some of the germplasm banks containing such genotype accessions (Schwember 2008). These germplasms also provide sources of genes conferring resistance to biotic and abiotic stresses, in addition to the most useful and valuable nutritional quality traits.

The most critical limitation related to genotype competency during genetic manipulation in soybean has been described by many authors as genotype specificity. This term ‘genotype specificity’ remains synonymous with the poor transformation rates that are currently being reported by many laboratories around the world dealing with soybean transformation. This phenomenon has been extremely inhibitory to genetic transformation but less insightful in distinguishing elite genotypes from those that are still not amenable to transformation. The raised points clearly indicate that work must continue on genotype characterisation to obtain basic information that will enable the identification of genetically different groups of soybean lines that are regenerable. Many monocots and legume crops that have high transformation rates showed genetic similarities when they were analysed for genetic manipulation. Displaying high genotype competency and regenerability in culture would permit significant levels of gene transfer and expression to take place.

However, this may not be surprising because legumes like soybeans are more self-pollinating than monocots, which have very high crossing rates. In soybean, the stigma of the pistil is completely covered by anthers of the stamen, making cross pollination nearly impossible. But a study reported higher genetic variation within a single population (Hs=0.315) and lower genetic divergence among other population than would be expected from an autogamous *Glycine soja* Sieb and Zucc (Fujita et al. 1997). Perhaps like in monocots, the first efficient transformation of soybean reported by Hinchee et al. (1988) should have revolutionised genetic transformation at a larger scale in most crop legumes, including cowpea and mung bean. The first soybean transformation should have provided insights on the physiological, metabolic and genomic processes involved during plant transformation. Furthermore, gene expression analysis should have also yielded comprehensive information to help in identifying genes responsible for genotype competency and regenerability traits found in certain elite cultivars like Peking and William.

6.2.2 Choice of Explants

Previous studies have asserted the utilisation of cotyledonary node explants for improved frequency of *in vitro* regeneration and the recovery of transformed shoots. Seldom, higher regeneration rates were achieved, coupled with a successful application of genetic transformation technology. About 75–100% shoot induction frequency was recorded in various cultivars of soybean (A6785, Bragg, Bunya, Fernside, Jack, MoonB1, Snowy and William) using cotyledonary node explants subcultured on Gamborg’s B5 basal medium supplemented with 1.67 mg/L benzyladenine (BAP) (Raza et al. 2017). An orthogonal design was used to assess the influence of these cotyledonary node explants, genotype and PGRs on shoot regeneration frequency using MSB_5 medium supplemented with 3.0 mg/L BAP, 0.2 mg/L indole-3-butyric acid (IBA) and 0.5 mg/L kinetin (KI) in soybean cultivar Hefeng 25. This study by Ma and Wu (2008) indicated that cotyledonary explants could regenerate shoots ranging from 30 to 35 per explant.

This culture optimisation did not only enhance shoot regeneration but also indicated that shoot proteins accumulated gradually and reached a peak at late shoot bud formation, just two months post-culture establishment. In a related study, Zhang et al. (2014) reported an optimised regeneration and transformation protocol using *A. tumefaciens* based on whole cotyledonary node explants established in soybean cultivar Huang 13. Soybean cultivar NARC-4 and NARC-7 both achieved 88 and 82% regeneration frequency also using a similar kind of explants. About 7.3 shoots per explant were induced on average (Zia et al. 2010). Nonetheless, variations in the number of shoots were attributed to several factors, including the genotype, hormonal combination, the type of media and most critically the explant type used. In contrast, some studies indicated that half-cotyledonary node explants could be exploited for direct micropropagation and transformation in soybean. An efficient plant regeneration requires explants capable of directly developing multiple shoots on a range of basal culture media, possibly devoid or with a lesser amount of growth regulators.

Cotyledonary nodes are often recommended as such kind of explants, which have thus far showed the capability of inducing high rates of multiple shoots when both cotyledons remain attached to the embryogenic axis to support shoot bud proliferation. This embryogenic axis or cotyledonary junctions currently used to make incisions for *A. tumefaciens* infection during transformation provide the infecting bacterial cell space to penetrate and adhere to the surfaces of the targeted cells. These explants provide the best source of totipotent cells and are the reason why the cotyledonary node system serves as the best method for plant transformation in modern biotechnology. Based on the findings made in many studies, the use of cotyledonary nodes as explants has been viewed as the most successful regeneration system for genetic transformation (Hinchee et al. 1988; Paz et al. 2006; Zhang et al. 2014). This approach has been considered as the most comparatively rapid and efficient method of regeneration via *de novo* organogenesis.

6.2.3 Co-Cultivation Conditions

Co-cultivation conditions form part of several major objectives in soybean transformation that researchers should seriously consider, in addition to the selection of amenable genotypes, competent explants, efficient selective processes, rapid regeneration and efficient recovery of putative transgenic plants. These objectives can be potentially achieved using a thoroughly optimised scheme that follows a series of steps as indicated previously in Chapter 4. The culture conditions of this transformation design scheme should have a clearly defined genetic engineering strategy that would enable the efficient and rapid transfer of T-DNA into targeted host tissues without causing severe senescence and deaths of explants. Currently, the purpose of many research teams and laboratories dealing with soybean transformation is to develop an improved technique for efficient *in vitro* regeneration and transgene expression.

However, these advancements would require an effective co-cultivation stage containing appropriate culture additives with proper incubation conditions to promote gene integration. In the past, soybeans have shown significant variability on *in vitro* growth and proliferation of shoots based on the photoperiod and temperature requirements (Jepleting 2015; Wawrzyniak et al. 2020). In tissue culture, significant efforts have to be made on the improvement of co-cultivation conditions involving soybean regeneration. Generally, these can be achieved as part of the overall modifications of a series of steps outlined in Chapter 4, particularly at the co-cultivation stage where gene transfer between explant tissues and *A. tumefaciens* must be effected, followed later by the isolation and determination of putative transgenic cells.

Any attempts to use *Agrobacterium* as a vehicle for the genetic manipulation of crop species should thoroughly determine the composition of the co-cultivation medium (with the inclusion of acetosyringone, cysteine and dithiothreitol, etc.), duration for this stage and the elimination of bacterial cells after co-cultivation. These factors were all discussed in greater detail in Chapter 4. As previously delineated, problems caused during explant-*Agrobacterium* interactions such as contamination, *Agrobacterium* overgrowth, explant tissue darkening, senescence and oxidative stress will pose serious challenges to the efforts at improving transformation frequencies. Co-cultivation serves as the most important and central stage of transformation that determines before *in vitro* regeneration whether transgenic events will possibly take place or not. This is the stage where most genetic transformations fail to take place. Furthermore, the levels of sensitivity of explants to *Agrobacterium* infection will be revealed in this stage, and this greatly influences transformation efficiency as alluded to by Carvalho et al. (2004).

6.2.4 Selection System

The transfer and expression of the T-DNA in host cells are very complex biological processes necessitating the activation of numerous genetic codes that are contained within the Ti-plasmids. But this process indicates that transgene expression relies on culture conditions, including other intrinsic and extrinsic factors that influence *vir* gene expression and plant host cell responses. Among these, selection regimes to distinguish between transformed and non-transformed cells serve as one of the most critical stages during transformation, especially post-infection and co-cultivation of explants with *A. tumefaciens*. There are several ways in which transgenic plants can be selected. In most reports, the selection system and co-cultivation medium serve as the most important factors determining the success of transformation and recovery of transgenic plants (Joyce et al. 2010). Delivered alongside the T-DNA are selectable marker genes which are a vital part of soybean transformation protocols. Selectable marker genes confer the ability for transformed plants to grow in the presence of a selective agent that is toxic to non-transformed cells or inhibitory to the growth of non-transformed plants.

Such toxic selective agents include antibiotics (e.g., kanamycin) and herbicides (e.g., glufosinate ammonium) that are applied to the medium after co-cultivation of explants with *Agrobacterium*. The genes that are frequently used to distinguish transformed plant tissues from those that are not transgenic are summarised in Table 6.1 (Mehrotra and Goyal 2010).

TABLE 6.1
Selectable marker gene systems used for the selection and identification of transgenic cells in plants

Marker system	Gene	Selective agent
Neomycin phosphotransferase	*nptII*	Kanamycin
Hygromycin phosphotransferase	*hpt*	Hygromycin B
Phosphinothricin acetyltransferase	*bar*	Phosphinothricin
Glyphosate oxidoreductase	*Gox*	Glyphosate
E-enolpyruvyl shikimate-3-phosphate	*EPSPS*	Glyphosate
Chloramphenicol reductase	*dbfr*	Methotrexate
Phosphomannose isomerase	*ManA*	D-Mannose

Agrobacterium-mediated transformation remains one of the only two predominantly employed methods of genetic engineering with biolistic mediated transformation. This approach mostly uses selectable marker genes highlighted in Table 6.1 combined with reporter genes such as *GUS/uidA*, *afp*, *LucZ* and *Luc* that use β-glucuronidase, green fluorescence protein, galactosidase and luciferase enzyme systems, respectively (Barampuram and Zhang 2011). These selectable and reporter genes enable the effective selection of cells/tissues carrying the gene of interest. In the current plant transformation systems, including *Agrobacterium*-mediated genetic transformation, these genes are co-delivered with the gene of interest to targeted host plant cells in order to select transgenic cells from non-transformed ones.

6.2.5 Genetic or Chimeric Effects

Unintended genetic or chimeric effects also cause major challenges when dealing with the genetic modification of plants. But these effects are predominantly associated with indirect genetic engineering methods rather than direct gene transfer. Any modifications that lead to abnormal effects or unintended genetic changes of regenerated plants which are different from traits of the parent plants are more common in tissue culture. This implies that not all regenerants are exact clones (calliclones, mericlones and protoclones) from callus, meristem and protoplast cultures of the parent plant species. Such variations may be due to epigenetic factors, other genetic alterations by mutations or somaclonal variations that are caused by physiological, genetic and biochemical changes during plant tissue culture. Somaclones refers to plantlets with variations coming out of tissue culture (Trigiano and Gray 2005) as also earlier described and identified by Larkin and Scowcroft (1981). Meanwhile, mutations refers to changes in DNA sequence resulting from genome copying mistakes, exposure to ionising radiation and hazardous chemicals or from infections by plant pathogens (McDonald and Linde 2002).

The kind of genetic effects or chimerism taking place in tissue culture occurs when regenerants contain portions of their tissue system carrying the gene of interest; meanwhile other sections of the plant remain untransformed. This has been a fairly common phenomenon leading to the reduction of transformation efficiencies in legume species such as chickpea and cowpea (Bhowmik et al. 2019). Even though the presence of chimeric genes limits the efficiency of recovering transgenic plants, very minimal efforts have been directed towards research and reporting on this issue, especially in dealing with chimerism as one of the factors influencing the rate of transformation in legumes. Perhaps this is also an indication of the diminishing interest in *Agrobacterium*-mediated genetic transformation in recalcitrant legumes due to its persistent difficulties. This non-transmission of genes across plant genome must be thoroughly investigated, rather than being reduced to the problem of low transformation efficacies, lack of reproducible protocols or poor *in vitro* rooting as it is currently purported (Popelka et al. 2004). Chimerism in plant engineering remains a challenge while this method is considered to be more precise and efficient in developing new cultivars than using conventional breeding methods for the genetic improvement of legume crops.

6.3 EMERGING FACTORS AFFECTING SOYBEAN TRANSFORMATION

The above discussion summarised the five primary factors influencing the *in vitro* regeneration of transformed soybean plants. These factors primarily affect the proliferation capacity of the *Agrobacterium*-infected tissues in a culture, efficiency of transgenesis and subsequently, the recovery of transgenic plants post-infection and co-cultivation stages. However, there are more neglected and emerging factors that are also known to negatively influence the amenability of soybean genotypes to *Agrobacterium*-mediated genetic transformation. Furthermore, their impact is also associated with the unintended and unwanted consequences of the primary factors previously discussed above. It is, however, clear that all of these factors combined result in a 'three-way interaction' that mediate the overall plant response to genetic improvement. Genotype responses vary according to the genetic constitution, amenability/susceptibility levels to *A. tumefaciens* infection and genetic fitness of plants in dealing with oxidative stress as well as tissue senescence induced during the transformation process.

In general, the adaptive flexibility of the genotype, its anatomical responses or differences of target tissues inherent to the plant genotype and the structural morphology of the explants/explant source all have serious impact on plant transformation. In addition to the longstanding factors affecting *Agrobacterium*-mediated transformation culture, factors presented below also have serious consequences if left unattended during culture protocol standardisation.

6.3.1 SEED QUALITY

Seed quality involves parameters such as the genetic and physical purity of the seeds. These are further accompanied by good physiological status, health and the exclusion of contaminating agents. Seeds with high physical quality have uniform sizes, weights and colour. The contamination of seeds by weeds or those that are disease mottled were reported to significantly reduce the physical purity and germination (Bishaw et al. 2012). Generally, the lack of physical quality or purity in seeds indirectly affects plant growth establishment during cultivation. Seeds are functional reproductive units and therefore, it is their true-to-type nature resembling the mother plants that helps them to achieve desired yield or resistance to biotic and abiotic stress factors. Soybeans, like other legume crops, require genetic improvement in order for farmers and consumers to reap the benefits from the higher amounts of carbohydrates, proteins and oils stored in the cotyledons. However, seed quality (i.e., seed viability and vigour) plays a critical role during plant establishment and it is more associated with germination, growth and yields (McDonald and Copeland 1997; Carera et al. 2011; Hagely et al. 2013).

The poor seed vigour or viability is frequently displayed by soybean genotypes wherein rapid and excessive moisture losses take place and are considered to be a clear indication of the reduced quality levels of a seed lot (Afrakhteh et al. 2013). Moreover, seed quality also has negative effects on seed utilisation as high energy sources and as propagules. The reproductive functionality of soybean seeds as propagules depends upon quality-related factors such as health, size, vigour and viability. Many researchers emphasise how seed quality directly influences successful plant establishment and productivity, but neglect its indirect role in *in vitro* plant regeneration and transformation. But like other pulses, soybean also remains recalcitrant, showing poor germination even under aseptic *in vitro* conditions.

Roos (1989) reported an increase on seed viability and longevity that translated into better performance by the adjustment of moisture, storage conditions and temperature. Providing a stable relative humidity and maintaining the seeds under 30°C demonstrated a positive logarithmic relation between seed viability and storage conditions. The deterioration of seeds in soybean takes place very rapidly, and this also has negative effects on the quality of explants sourced from such seeds. The scientific basis of the correlation between seed storage, germination and plant regeneration in soybean was also reported (Mangena and Mokwala 2019). Findings made in this study laid fundamental grounds in exploring and considering the effects of seed quality during genetic improvement via *A. tumefaciens*-mediated genetic transformation.

6.3.2 Seasonality

The effects of seasonality on plant response form part of a myriad of the most neglected factors in plant transformation or tissue culture as a whole. This involves changes occurring at the physiological or phenotypic states, that impact negatively on plant tissue culture responses. *In vitro* multiplication rates and regeneration can be very high during certain seasons since genotype effects and growth conditions may influence culture responses. Plant tissue culture is designed to provide year-round plant production, with the propagation easily scheduled according to the market demand (Barnard 1997; Bewley 1997; Hartmann et al. 2014). This technique largely allows for the manoeuvre of production and storage facilities, making it possible to produce plants without genetic predetermination interference. Techniques like nodal culture or shoot tip culture may be easily established if the explants are collected from plants during relatively active growth.

For many cultures, especially those involving woody plants, the type of explant tissues utilised may often be at the dormant or resting stage which is also seasonal (Jiang et al. 2011). Such varied explant parts may also be difficult to sterilise, like the explants collected from expanding growths which are easier to disinfect than those collected directly from new outgrowths. A strong seasonal variation in plant regeneration frequency and transformation was observed from two barley (*Hordeum vulgare*) cultivars Golden Promise and Solome. Scutella immature embryos were bombarded with a plasmid pAHC20 containing the *bar* selectable marker gene conferring tolerance to bialaphos. Scutella transformation in cv. Salome was high between March and April, and then dramatically decreased in May to December (Sharma et al. 2005).

Other studies that investigated the influence of different seasons on *in vitro* culture establishment involved shoot proliferation in *Bambusa nutans* Wall. Ex. Munro (Mudoi et al. 2014) and callus induction and shoot proliferation as well as somatic embryogenesis from anther culture of *Hevea bransiliensis* (Srichuay et al. 2014). In soybean, the effect of seasonality on culture establishment and transformation has not yet been thoroughly explored and reported.

6.3.3 Sterilisation Techniques

The method of sterilisation chosen can also influence the establishment and efficiency of transformation cultures. The initiation and aseptic establishment of pathogen eradicated cultures is among the main goals of *in vitro*-based techniques. This stage is very important since the presence of microbial contaminants can adversely affect the responsiveness of plant tissue explants used for *Agrobacterium*-mediated genetic transformation. Culture contamination can adversely affect *in vitro* seed germination, shoot induction and elongation/rooting stages. Any negative influence of contamination will reduce the recovery rates of transformed plantlets. It is essential that whenever possible, the explant source is indexed for the presence of internal viral contaminations even before being prepared for any type of *in vitro* culture propagation and genetic transformation. Indexing is more preferable when dealing with fastidious bacteria, intracellular and intercellular endophytes which often go undetected during clonal propagation via tissue culture (Clemente and Cahoon 2009). For detailed fundamental information on approved methods for the indexing of plants for pathogen eradication see Chapter 28, pages 321–331 by Trigiano and Gray (2005).

Factors influencing aseptic culture establishment and responsiveness may include seasons and the type and size of explants used, together with the production of polyphenol oxidation. These factors cause the differences in response capacities of explants for *in vitro* culture establishments. Many transformation cultures are adversely affected by polyphenol oxidation products and *A. tumefaciens* overgrowth which are capable of killing the explants faster than any externally/internally introduced microbial contaminants that can be easily controlled using broad-spectrum antibiotics. Polyphenol accumulations can severely blacken the explants exposing them to oxidative stress, and interfering with nutrient availability to growing tissues in the medium (Mangena 2021). The effects of *Agrobacterium* overgrowth during plant transformation cultures were comprehensively covered in the previous chapters. However, information is very scarce on this topic, including the formation of vitropaths or those that may positively influence plant growth and development through medium acidification affecting the release and availability of PGRs for cultured plant tissues (Cassells 2001; Miransari and Smith 2014).

6.3.4 Plant Growth Regulators

Although the use of PGRs such as auxins, cytokinins and gibberellins is very common and mandatory in plant tissue culture and genetic transformation, successes in the regeneration of transformed plants can also be achieved through the manipulation of the quality of explants, cultivation periods and efficient control of conditions such as light, temperature, humidity and carbon source (Gallardo et al. 2001; Stefanova et al. 2011). Of course, exogenously applied PGRs play an important role in causing desirable histologenical and morphogenical changes in a tissue culture depending on the type and concentrations used. But they may also result in abnormal anatomical, morphological and physiological changes of plantlets developed *in vitro*. Such abnormalities may include the lack of epicuticular wax, thinner mesophylls and low stomatal conductance, in addition to the poor structural and functional characteristics of leaves, stems and roots (Brainerd and Fuchigami 1981; Hazarika 2006; Magyar-Tabori et al. 2010). All of these changes have negative effects on the performance of *in vitro* regenerated plantlets, either transformed or non-transformed, especially during acclimatisation stages. Presently, there is limited literature about the role of transformation techniques and conditions in structural and functional changes taking place in major organs of transformed plantlets as a result of moderate or excessive applications of PGRs.

6.3.5 Antibiotics

Both the aminoglycoside and beta-lactam broad-spectrum antibiotics are important to the genetic transformation of soybean and other crops. Aminoglycosides are broad-spectrum antibiotics that target many types of Gram-negative bacteria, with the exclusion of a few such as *Agrobacterium tumefaciens*. Findings made in our previous studies indicated that these antibiotics may only be effective against genetically engineered strains of *A. tumefaciens* when used in very high concentrations. Amounts ranging between 300 and 500 mg/L were tested for *Agrobacterium* strain EHA101.1 carrying the vector pTF101.1 and another cell line with the vector ΩPKY. According to the previous findings, the addition of hygromycin, streptomycin, tetracycline, spectinomycin and rifampicin to a culture medium at high concentrations effectively suppressed the growth of *A. tumefaciens* through the agar diffusion assay (Mangena 2015). Tetracycline, spectinomycin, rifampicin and kanamycin are all routinely used for soybean transformation as selective agents in the growth of this bacterium, as well as for the selection and identification of transgenic plants.

Although these antibiotics are effectively used during genetic transformation, some evidence pointed out some undesirable levels of toxicity in plant tissues. These antibiotics were found to cause early tissue senescence in soybean cotyledonary node explants as was similarly observed in *Nicotiana plumbaginifolia* (Pollock et al. 1983). On the other side, beta-lactams were also found to reduce the frequency of shoot proliferation. Tang et al. (2000) and Alsheikh et al. (2002) reported severe negative effects when an increased concentration of these antibiotics was used as bactericides of choice for the elimination of *A. tumefaciens* post-explant infection and co-cultivation. β-lactams antibiotics contain a beta-lactam ring in their molecular structure (3-carbon and 1-nitrogen ring) and act against many Gram-positive and Gram-negative bacteria (Mora-Ochomogo and Lohans 2001). These antibiotics are usually safe and effective against *A. tumefaciens*, and are well tolerated by plant tissue explants during *in vitro* cultures.

6.3.6 Oxidative Stress

As discussed previously, oxidative stress is caused by the imbalances between the induction and accumulation of reactive oxygen species (ROS) and secondary metabolites. Reactive oxygen species commonly cause oxidative stress during plant exposure to biotic and abiotic stress. However, our findings suggest that ROS are also induced as a response to explant infection with *A. tumefaciens* during *in vitro* transformation. The establishment of an efficient protocol in soybean relies upon effective interactions between explants and *Agrobacterium*, particularly an interaction that is marked by very minimal ROS or exudates accumulation that may cause excessive oxidative stress. In fact, tissue oxidation could either be caused by the production of ROS or exudates produced when plant organs are injured during explant preparation and infection. The kind of exudates produced may be of allelochemic nature, which are secondary metabolites that function to counterbalance biotic and abiotic stress under natural conditions (Crowe and Crowe 1992).

The overexpression of these polyphenols usually causes browning or blackening of explant tissues and the culture medium, as well possible stimulation to regulate and inhibit the growth of transforming bacterial cells. Oxidative browning is a common challenge in tissue culture and *in vitro*-based genetic transformation of plants. Generally, it results in reduced growth of target cells infected with *Agrobacterium*, lowers transformation rates and ultimately kills the explants. Unlike during the exposure of plants to environmental stress, the frequency of plant tissue culture-based oxidation varies according to species, cultivar, physiological state and type of explants used (Gallardo et al. 2001; Jones and Saxena 2013). The underlying problem during plant transformation is that the overexpression of these secondary metabolites as a result of tissue wounding during explant preparation or defence response triggered by the invasion of bacterial cells into host cells get oxidised causing oxidative burst.

6.4 SUMMARY

The widespread accumulation of reactive oxygen species and harmful secondary metabolites during transformation continues to increasingly be one of the causes of reduced transformation efficiencies. Often it is the simultaneous occurrence of these ROS and polyphenols oxidation that reduces the totipotency of targeted plant regenerating cells, followed by *Agrobacterium* overgrowth and the genotype factor. Even at lower concentrations, polyphenolic oxidations can be harmful to explants, interfering with the transforming bacterial cells during co-cultivation. Plants naturally produce alkaloids, flavonoids, tannins, terpenes, resins and anti-inflammatory agents. Resins, tannins, flavonoids and alkaloids form part of major groups of secondary compounds found across plant species, including species of the soybean (Fabaceae) family (Compean and Ynalvez 2014). Previous revelations of certain secondary metabolites as potential antibacterial agents may, therefore, be responsible for the reduction of transgene transfer and expression taking place in wounded host plant tissues during genetic transformation using *A. tumefaciens*.

For several years, there has been increased interest in the co-cultivation conditions of explants with *A. tumefaciens* carrying plasmid vectors that contain the gene of interest. It was reported that increased infection and co-cultivation periods improved rates of transformation. These effects were further complemented by the addition of organic supplements such as dithiothreitol, L-cysteine, acetosyringone and PGRs like BAP, gibberellic acid and ascorbic acid (Somers et al. 2003; Paz et al. 2006; Ma and Wu 2008; Verma et al. 2014). A combinational use of these developmental stimuli triggers favourable proliferation of cells, including those that may contain the gene of interest. PGRs are involved essentially in all aspects of plant growth and development. BAP for instance controls cell division and cytokinesis during shoot induction. It also mediates physiological processes such as delaying tissue senescence and shoot multiplications. There is substantial evidence found in reports cited in this chapter which suggest that PGRs and other organic supplements used in *in vitro* transformation primarily act in transgene expression in host plant tissues.

Although *Agrobacterium* has the ability to shift the balance of metabolites in infected plant tissues, the effects of endogenous growth-promoting metabolites may dramatically affect these abilities and then lead to *Agrobacterium*-induced tissue senescence. This serves as a clear indication that effective optimisation and separation of transformation stages are required. An effective disinfecting method is required to minimise the role of bacterial cell overgrowths in plant tissue senescence, especially because of the increased prevalence of antibiotic resistance emerging in plant tissue culture from the extensive use of antimicrobials. This renders the current β-lactam antibiotics insufficient for controlling the growth of contaminating bacteria (including *A. tumefaciens* overgrowth) and fungi during transformation cultures. A comprehensive study of the role of antibiotics (cefotaxime, vancomycin and carbenicillin) and *Agrobacterium* response in threshold survival levels in *Chrysanthemum* and *Nicotiana* tissues was published by da Silva and Fukai (2001).

6.5 ABBREVIATIONS

AMGRC	Australian *Medicago* Genetic Resources Centre
ATFCC	Australian Temperate Field Crops Collections
BAP	6-benzylaminopurine
CIAT	International Centre for Tropical Agriculture
DNA	Deoxyribonucleic acid
GUS	β-Glucuronidase
IBA	indole-3-butyric acid
ICARDA	International Centre for Agricultural Research in the Dry Areas
ICGR-CAAS	Institute of Crop Germplasm Resources – Chinese Academy of Agricultural Sciences
ICRISAT	International Crops Research Institute for the Semi-Arid Tropics
IITA	International Institute for Tropical Agriculture
KI	Kinetin
Luc	Luciferase
MSB_5	Murashige and Skoog Gamborg's B_5 medium
PGRs	Plant growth regulators
ROS	Reactive oxygen species
T-DNA	Transferred deoxyribonucleic acid

REFERENCES

Afrakhteh S, Frahmandfar E, Hamidi A and Ramandi HD. (2013). Evaluation of growth characteristics and seedling vigor in two cultivars of soybean dried under different temperature and fluidized bed dryer. *International Journal of Agriculture and Crop Sciences* 5(21), 2537–2544.

Alsheikh MK, Suso HP, Robson M and Battey NH. (2002). Appropriate choice of antibiotics and *Agrobacterium* strain improves transformation of antibiotic sensitive Fragaria vesca and F.V semperflorens. *Plant Cell Reports* 20, 1173–1180.

Barampuram S and Zhang L. (2011). Recent advances in plant transformation. In Birchler J (eds), *Plant Chromosome Engineering: Methods in Molecular Biology (Methods and Protocols)*, Vol 701. Humana Press, Totowa, New Jersey. pp. 1–35.

Bernard RL. (1997). Two major genes for time of flowering and maturity in soybeans. *Crop Science* 11(2), 242–244.

Bewley JD. (1997). Seed germination and dormancy. *The Plant Cell* 9, 1055–1066.

Bhowmik SSD, Cheng AY, Long H, Tan GZH, Hoang TML, Karbaschi MR, Williams B, Higgins TJV and Mundree SG. (2019). Robust genetic transformation system to obtain non-chimeric transgenic chickpea. *Frontiers in Plant Science* 10(524), 1–14.

Bishaw Z, Straik PC and van Gastel AJG. (2012). Farmers' seed sources and quality: 1. Physical and physiological quality. *Journal of Crop Improvement* 26(5), 655–692.

Brainerd KE and Fuchigami LH. (1981). Acclimatization of aseptically cultured apple plants to low relative humidity. *Journal of the American Society for Horticultural Science* 1064, 515–518.

Carrera CS, Reynoso CM, Funes GJ, Martinez MJ, Dardanelli J and Resnik SL. (2011). Amino acid composition of soybean seeds as affected by climatic variables. *Pesquisa Agropecuaria Brasileira* 46(12), 1579–1587.

Carvalho CHS, Zehr UB, Gunaratna N, Anderson J, Knonowicz HH, Hodges TK and Axtell JD. (2004). *Agrobacterium*-mediated transformation of sorghum: Factors that affect transformation efficiency. *Genetics and Molecular Biology* 27(2), 259–269.

Cassells AC. (2001). Contamination and its impact in tissue culture. *Acta Horticulturae* 560, 353–359.

Clemente TE and Cahoon EB. (2009). Soybean oil: Genetic approaches for modification of functionality and total content. *Plant Physiology* 151(3), 1030–1040.

Compean KL and Ynalvez RA. (2014). Antimicrobial activity of plant secondary metabolites: A review. *Research Journal of Medicinal Plant* 8(5), 204–213.

Crowe JH and Crowe LM. (1992). Membrane integrity in anhydrobiotic organism: Toward a mechanism for stabilising dry seeds. In Somero GN, Osmond CB and Bolis CL (eds), *Water and Life*. Springer-Verlag, Berlin. pp. 87–103.

da Silva JAT and Fukai S. (2001). The impact of carbenicillin, cefotaxime and vancomycin on *Chrysanthemum* and tobacco TLC morphogenesis and *Agrobacterium* growth. *Journal of Applied Horticulture* 3(1), 3–12.

Fujita R, Ohara M, Okazaki K and Shimamoto Y. (1997). The extent of natural cross-pollination in wild soybean (*Glycine soja*). *Journal of Heredity* 88, 124–128.

Gallardo K, Jub C, Groot SPC, Peype M, Demol H, Vandekerchore J and Job D. (2001). Proteomics of *Arabidopsis* seed germination. A comparative study of wildtype and gibberellin deficient seeds. *Plant Physiology* 129, 823–837.

Hagely KB, Palmquist D and Bilyeu KD. (2013). Classification of distinct seed carbohydrates profiles in soybean. *Journal of Agricultural and Food Chemistry* 61(5), 1105–1111.

Hartmann HT, Kester DE, Daries FT and Geneve RL. (2014). *Plant Propagation: Principles and Practices*. Person's Education, Inc., Edinburgh Gate, Harlow p. 147.

Hazarika BN. (2006). Morpho-physiological disorders in *in vitro* culture of plants. *Scientia Horticulturae* 108, 105–120.

Hinchee MAM, Connor-Ward DV, Newell CA, McConnell RE, Sato SJ, Gasser CS, Fischhoff DA, Re DB, Fraley RT and Horsch RB. (1988). Production of transgenic soybean plants using *Agrobacterium*-mediated DNA transfer. *Biotechnology* 6, 915–922.

Jacob C, Carrasco B and Schwember AR. (2016). Advances in breeding and biotechnology of legume crops. *Plant Cell Tissue and Organ Culture* 127, 561–584.

Jepleting CG. *Seed Quality of Soybean (Glycine max L. Merrill) Genotypes under Varying Storage and Priming Methods, Mother Plant Nutrient Profile and Agroecologies in Kenya*. PhD Thesis, Kenyatta University, Kenya. 2015.

Jiang Y, Wu C, Zhang L, Hu P, Hou W, Zu W and Han T. (2011). Long-day effects on the terminal inflorescence development of a photoperiod sensitive soybean [*Glycine max* (L.) Merr] variety. *Plant Science* 180(3), 504–510.

Jones AMP and Saxena PK. (2013). Inhibition of phenylpropanoid biosynthesis in *Artemisia annua* L: A Novel approach to reduce oxidative browning in plant tissue culture. *PLoS ONE* 8(10) e76802, 1–13.

Joyce P, Kuwahata M, Turner N and Lakshmanan P. (2010). Selection system and co-cultivation medium are important determinants of Agrobacterium-mediated transformation of sugarcane. *Plant Cell Reports* 29(2), 173–783.

Larkin PJ and Scowcroft WR. (1981). Somaclonal variation- A novel source of variability from cell cultures for plant improvement. *Theoretical and Applied Genetics* 60, 197–214.

Liu K. (1997). Chemistry and nutritional value of soybean components. In Liu K (ed), *Soybeans: Chemistry, Technology and Utilisation*. Springer Boston. pp. 25–113.

Ma X-H and Wu T-L. (2008). Rapid and efficient regeneration in soybean [*Glycine max* (L.) Merrill] from whole cotyledonary node explant. *Acta Physiologia Plantarum* 30, 209–216.

Magyar-Tábori K, Dobránszki J, Teixeira da Silva JA, Bulley SM and Hudák I. (2010). The role of cytokinins in shoot organogenesis in apple. *Plant Cell, Tissue and Organ Culture* 101, 251–267.

Mangena P. 2015 *Oryza Cystatin-1 Based Genetic Transformation in Soybean for Drought Tolerance.* Masters Thesis, University of Limpopo, South African.

Mangena P. (2021). Effect of Agrobacterium co-cultivation stage on explant response for subsequent genetic transformation in Soybean (*Glycine max* (L.) Merr.). *Plant Science Today* 8(4), 905–911.

Mangena P. *An Investigation on the Cause of Recalcitrance to Genetic Transformation in Soybean, Glycine max (L.) Merrill.* PhD Thesis, University of Limpopo, South Africa. 2019.

Mangena P, Mokwala PW. The influence of seed viability on the germination and *in vitro* multiple shoot regeneration of soybean (*Glycine max* L.). *Agriculture* 9(2), 35, 1–12.

McDonald BA and Linde C. (2002). Pathogen population genetics, evolutionary potential, and durable resistance. *Annual Review of Phytopathology* 40, 349–379.

McDonald MB and Copeland LO. (1997). Seed quality and performance. In McDonald MB and Copeland LO (eds), *Seed Production.* Springer, Boston. pp. 19–28.

Mehrotra S and Goyal U. (2010). Agrobacterium-mediated alien gene transfer biofabricates designer plants. In Watson RR and Preedy VR (eds), *Genetically Modified Organisms in Food Production, Safety Regulation and Public Health.* Academic Press, London. pp. 63–73.

Miransari M and Smith DL. (2014). Plant hormones and seed germination. *Environmental and Experimental Botany* 99, 110–121.

Mora-Ochomogo M and Lohans CT. (2001). β-lactam antibiotic targets and resistance mechanisms: From covalent inhibitors to substrates. *RSC Medicinal Chemistry* 12, 1623–1639.

Mudoi KD, Saikia SP and Borthakur M. (2014). Effect of nodal positions, seasonal venations, shoot clump and growth regulators on micropropagation of commercially important bamboo, *Bambusa nutans* Wall. Ex. Muntro. *African Journal of Biotechnology* 13(19), 1961–1972.

Paz MM, Martinez JC, Kalvig AB, Fonger TM and Wang K. (2006). Improved cotyledonary-node method using an alternative explant derived from mature seed for efficient *Agrobacterium*-mediated soybean transformation. *Plant Cell Reports* 25, 206–213.

Pollock K, Barfield DG and Sheilds R. (1983). The toxicity of antibiotics to plant cell culture. Plant Cell Reports 2, 36–39.

Popelka JC, Terryn N and Higgins T. (2004). Gene technology for grain legumes can it contribute to the food challenge in developing countries? *Plant Science* 2, 195–206.

Raza G, Singh MB and Bhella PL. (2017). *In vitro* plant regeneration from commercial cultivars of soybean. *BioMed Research International* 7379693, 1 9.

Roos EE. (1989). Long-term seed storage. In Janick J (ed), *Plant Breeding Reviews: The National Plant Germplasm System of the United States*, Vol 7. Timber Press Inc., Portland, Oregon. pp. 129–158.

Schwember AR. (2008). An update on genetically modified crops. *Ciencia e Investigacion AGRARIA* 35, 231–50.

Sharma VK, Hansch R, Mendel RR and Schulze J. (2005). Seasonal effect on tissue culture response and plant regeneration frequency from non-bombarded and bombarded immature scutella of barley (*Hordeum vulgare*) harvested from controlled environment. *Plant Cell, Tissue and Organ Culture* 81, 19–26.

Somers DA, Samac DA and Olhoft PM. (2003). Recent advances in legume transformation. *Plant Physiology* 131, 92–899.

Srichuay W, Kalawong S and Te-chato S. (2014). Effect of seasonal collection on callus induction, proliferation and somatic embryogenesis from anther culture of *Hevea bransiliensis* Muell Arg. *African Journal of Biotechnology* 13(35), 3560–3566.

Stefanova MA, Koleva DP, Ganeva TG and Dimitrova MA. (2011). Effect of plant growth regulators on the regeneration of *in vitro*-propagated *Lamium album* L. plants. *Journal of Pharmacy Research* 4(7), 1982–1985.

Tang H, Ren Z and Krczal G. (2000). An evaluation of antibiotics for the elimination of *Agrobacterium tumefaciens* from walnut somatic embryos and for the effect on the proliferation of somatic embryos and regeneration of transgenic plants. *Plant Cell Reports* 19, 881–887.

Trigiano RN and Gray DJ. (2005). *Plant Development and Biotechnology*. CRC Press, Boca Raton. pp. 301–308.

Verma K, Saini R and Rani A. (2014). Recent advances in the regeneration and genetic transformation of soybean. *Journal of Innovative Biology* 1, 15–26.

Wawrzyniak MK, Michalak M and Chmielarz P. (2020). Effects of different condition of storage on seed variability and seedling growth of six European wild fruit woody plants. *Annals of Forest Science* 77(58), 1–20.

Zhang F, Chen C, Ge H, Liu J, Luo Y, Liu K, Chen L, Xu K, Zhang Y, Tan G and Li C. (2014). Efficient soybean regeneration and *Agrobacterium*-mediated transformation using a whole cotyledonary node as an explant. *Biotechnology and Applied Biochemistry* 61(5), 620–625.

Zia M, Rizvi ZF, ur-Rehman R and Chaudhary MF. (2010). Micropropagation of two Pakistani soybean (*Glycine max* L.) cultivar from cotyledonary nodes. *Spanish Journal of Agricultural Research* 8(2), 448–453.

7 Alternative Techniques for Genetic Manipulations in Soybean

7.1 INTRODUCTION

Soybean serves as one of the most important legumes that continues to be used for various food and health-promoting functions after the suitable processing and manufacturing of soy-based products. Aspects such as growth, yields, physiological stress and genetic manipulation remain very significant for the enhancement of its cultivation and consumption. Genetic improvement is also done to explore the development of new stress-resistant varieties, new soy-products and to potentially diversify available genetic resources. Future research focussing on the physiological response and genetic improvements of soybean needs to be prioritised to improve its utilisation and nutritional grain qualities. A number of genetic improvement techniques are available that are used to effect the genomic changes required. Some of these tools serve as the most economic and highly efficient methods of genetic engineering discovered in the history of biotechnology.

Such methods hold the potential and promise to efficiently regenerate transgenic plants, especially for recalcitrant legume crops. The purpose of this chapter is to provide a brief discussion on some of the other alternative techniques that can be used to advance soybean growth and yield, in addition to *Agrobacterium tumefaciens*-mediated genetic transformation. The discussions are generalised and intended to provide a comprehensive view of the impact of these technologies on soybeans, as well as legumes in general. Included in this chapter are (1) a brief overview on the technical aspect and functions of the technique, (2) a brief survey of the application in soybeans and other crops and (3) lastly the status and use of these techniques including *A. rhizogenes* for the genetic improvement of plants.

7.2 ALTERNATIVE TECHNIQUES FOR GENETIC IMPROVEMENT

The production of stable transgenic soybeans continues to pose major challenges as many researchers believe that factors such as the low recovery rates of genetically altered plants are due to genotype specificity and the formation of chimeras in a culture that form part of a myriad of encountered problems. A variety of explants and culture conditions remain tested and optimised for the development of a routine

DOI: 10.1201/b22829-7

transformation protocol as discussed in the previous chapter and according to the reports by Olhoft and Somers (2001), Yan et al. (2000), Ko et al. (2004) and Zeng et al. (2004). These studies concurred that advancing *in vitro* transformation should harness the morphogenetic potentiality of embryogenic cells of the explants. The cells are predominantly found on immature and mature cotyledonary nodes, and could be used for rapid plant regeneration without the formation of chimeras if culture conditions are well standardised. On the other hand, other genetic improvement techniques such as particle bombardment, liposome-mediated DNA delivery, mutagenesis and CRISPR-Cas can be used without the many hurdles faced during *in vitro* transformation (refer to Chapter 6).

Yan et al. (2000) reported that *in vitro* transformations have been found to be more prone to the problem of somaclonal variation that leads to the formation of chimeras and subsequently, false-positive identification of transgenic shoots than under *in vivo* conditions. On the other hand, non-tissue culture genetic transformation methods were reported to have worked successfully for the genetic manipulation of some dicots, *Arabidopsis*, cotton and many cereal crops, but not for legume crops most of which remain recalcitrant. Therefore, the lack of a feasible, reliable and highly reproducible genetic transformation system using *A. tumefaciens*, particularly, under both *in vitro* and *in vivo* conditions for the improvement of soybeans or other dicot plants stimulates routine testing of direct gene transfer procedures which do involve the use of biological vectors. Such direct gene transfer methods that rely upon the delivery of naked DNA molecules into host plant cells can be broadly divided into (1) physical gene transfer methods that include electroporation, microinjection and particle bombardment, and (2) chemical gene transfer methods such as calcium phosphate precipitation, diethyl amino ethyl (DEAE) dextran-mediated DNA transfer and polyethylene glycol (PEG)-mediated DNA transfer.

7.2.1 Particle Bombardment/Biolistic Method

A simplified protocol of a tissue culture-based particle bombardment genetic transformation is summarised in Figure 7.1. Particle bombardment, also known as biolistics or the gene gun method, entails the acceleration of high-density DNA-coated carrier particles (of approximately two microns in diameter) passed through plant cells and leaving the foreign DNA inside bombarded tissues (Rivera et al. 2012). The gene gun method was established in 1987 at Cornell University for the genetic engineering of single or organised cells in cereals. However, apart from the cereals, major crops like soybean and cotton have been among some of the most popular species that were genetically modified using this technique. Particle bombardment is a very complex and expensive technique. It employs high-velocity metal particles for delivering active DNA segments into host tissues, which is a straightforward tissue engineering protocol.

The procedure eliminates barriers involved in a number of genetic transformation systems, such as tissue culture-induced variations, genotype specificity, intensive laboratory protocols and time factors in the recovery of transgenic plantlets (Christou 1994). The comparison between *Agrobacterium* and particle bombardment in terms

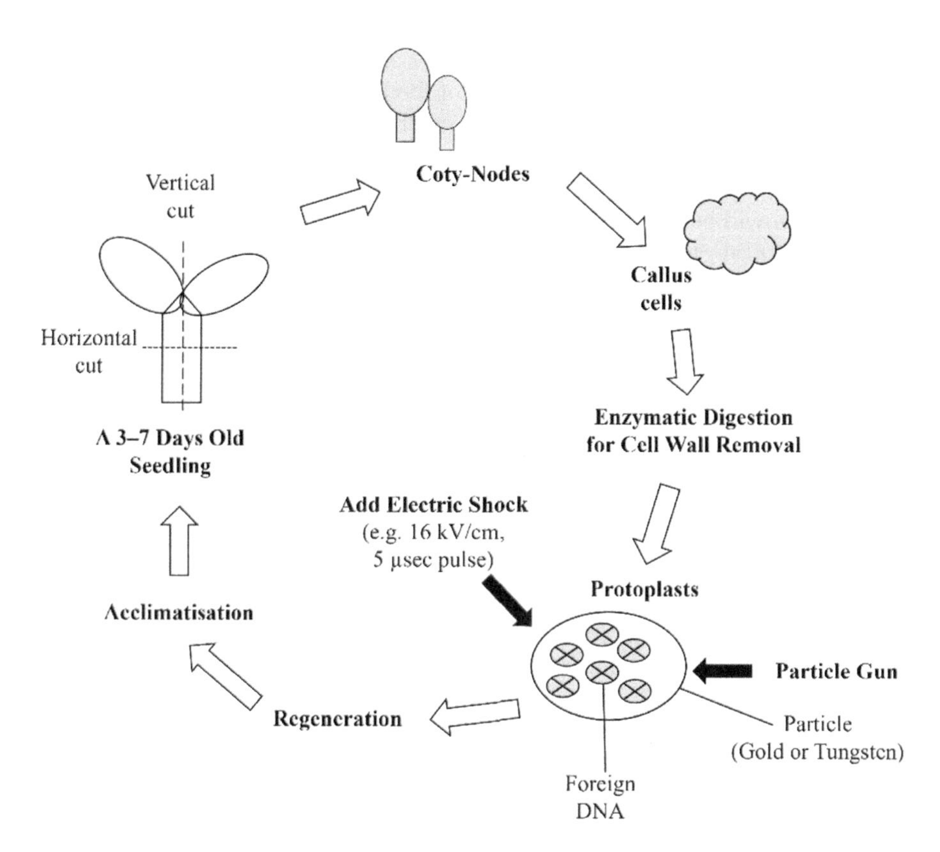

FIGURE 7.1 Schematic illustration of biolistic/particle bombardment procedure for crop improvement.

of transformation efficiency, transgene copy number, expression, inheritance and physical structure of the transgenic loci using fluorescence *in situ* hybridisation was investigated (Rivera et al. 2012). In general, the use of *Agrobacterium tumefaciens* offers significant advantages over biolistics, especially in terms of costs of operation. Nevertheless, organised or single somatic cells, undifferentiated meristematic cells, embryos (from germ or somatic cells), callus cells and protoplasts could be used as targeted 'explants' for genetic transformation in both tools (Figure 7.1). As such, this approach can be successfully employed for both nuclear and chloroplast transformation (Boynton et al. 1988).

7.2.2 Liposome-Mediated Transfection

Another tool used to inject foreign DNA fragments coding for desired proteins, conceivably conferring tolerance to biotic and abiotic stress, is liposome-mediated gene transfer. Deshayes and colleagues first reported this method, in which plasmid-loaded liposomes were fused with plant mesophyll protoplasts using polyethyleneglycol

(PEG) treatment (Deshayes et al. 1985). This approach makes use of positively charged liposomes (lipids) for the transfer of foreign genes into hosts. In this case, the gene of interest is encapsulated in the spherical lipid bilayer of the cell membrane. The procedure takes advantage of the favourable interaction between negatively charged DNA and positively charged liposomes. In instances where DNA is transferred, it will be free to recombine and integrate into the host genome following a successful invagination through endocytosis. A successful transfection using liposome-mediated gene transfer was previously reported in tobacco (Dashayes et al. 1985), potato and wheat (Behrooz et al. 2008).

Although inexpensive, liposome transfection is more widely used in mammalian cells than plant cells and tissues for DNA or protein delivery into host eukaryotic cells. This is done through the liposome polymer transfection complexes (LPTCs) which are usually used to condense transferred DNA and protect it against digestion by deoxyribonuclease enzymes (DNase I). DNase I are endonucleases that are mostly found in human genome used to cleave DNA by binding tightly to the minor grooves and sugar-phosphate backbone of both DNA double helix strands (Shiokawa and Tanuma 2001). However, soybean oil-based liposome-polymer transfection complexes were used by Lin et al. (2012) as a co-delivery system for DNA and vaccine subunits in mouse embryo fibroblast cell line Balb/3T3 cultured in Dulbecco's Modified Eagle Medium with 10% fetal bovine serum and 1% prostate-specific antigen.

7.2.3 Fibre-Mediated DNA Delivery

One of the most recently developed genetic transformation procedures includes silicon carbide fibre-mediated gene transfer. This method has been used to deliver foreign DNA in plant cells using plant tissue culture and the selection of transformed plants using genetic markers. Plant tissues that contain identifiable transferred marker genes are proliferated and regenerated further into transgenic plants using *in vitro* cultures. *Zea mays* (Black Mexican sweet maize) and *Nicotiana tabacum* (tobacco) were some of the first plants to be genetically engineered using the silicon carbide fibre-gene transfer in the presence of plasmid DNA containing a *GUS* reporter gene (Kaeppler et al. 1990). The transformed cells of tobacco and maize were vortexed and visualised using a scanning microscope to determine transgenic cells.

GUS expression of vortexed embryo cells, followed by transformed callus, was also identified in wheat (Serik et al. 1996). Comparatively, there are thus far no reports of transformation that obtained more than 6.88% efficiency (Hassan et al. 2021) in any crop, including soybean, that can be compared with the rates achieved using *Agrobacterium tumefaciens*. Rech et al. (2008) earlier reported 9, 2.7 and 0.55% efficient recovery of transgenic soybean, common bean and cotton plants, respectively. This is significantly higher than transformation efficiencies obtained using fibre gene transfers. However, this system offers very rapid transfections, low operation costs, simplified transformation procedures and proven effective means of recombination across various genotypes as some of the benefits of using this tool. Only low transformation efficiencies, severe tissue damage and toxic effects of the fibre are common and widely reported transformation problems (Behrooz et al. 2008).

7.2.4 Laser-Induced Genetic Improvement

The laser-induced intracellular transfer of foreign genomic materials involves the use of laser beams to puncture plant cell membranes to transfer exogenous DNA into the cell cytoplasm near the nucleus (Meacham et al. 2014). This technique is practised in both agricultural biotechnology and medical applications. But its application in mammalian cells and tissues was discouraged since it causes severe collateral tissue damage (Doukas and Kollias 2004). However, the importance of this laser application is its ability to delivery exogenous DNA fragments to mammalian and plant cells and tissues. This method was first introduced by Tsukakoshi and colleagues in 1984, in the delivery of exogenous gene materials into living cells using ND:YAG laser with 355 nm wavelength for 5 nanoseconds. In contrast to chemical gene transfer methods that are based on the phagocytosis of DNA precipitation, the laser method employs ultraviolet optical needles to prick cells.

It penetrates a self-healing hole of submicrometric diameter in the cell membrane to allow the foreign DNA contained in the medium passage into the targeted cell. In plants, laser-mediated gene transfer was reported to achieve high efficiency of *in vitro* and *in vivo* transformation if laser beams are focussed slightly within the cell near the nucleus as opposed to anywhere in the cytoplasm. Although there are numerous reports on the incorporation of exogenous DNA in plants using this technique, no reports have come to the fore regarding genetic improvement of soybean using laser radiation. Recently, a study illustrating the use of laser light for the delivery of double-stranded DNA (dsDNA) in *Citrus* leaves was reported by Killing et al. (2021).

7.2.5 *In-Planta* Transformations

There are several *in-planta* physical methods of genetic transformation that have been described in the past three decades. Most of them failed the test of time because they were not reproducible and the results obtained led to ambiguous interpretations. Methods such as *Agro*-injection of cotyledons, vacuum infiltration and microinjection of reproductive organs form part of the widely tested protocols applied in many species from *Arabidopsis* to legume crops such as soybean. All of these techniques bypass *in vitro* plant cell and tissue culture which allows for the regeneration of transgenic plants from a single cell. However, like *in vitro* transformation, *in-planta* genetic transformation still faces several challenges and has not led to more desirable heritable modifications as sought by many researchers. One of the most widely used *in-planta* techniques is the microprojection of protoplasts.

This approach uses low melting point (LMP) agarose for holding and culturing DNA microinjected protoplasts (Behrooz et al. 2008). Thus far, the most successful microspore-derived embryoids were developed in *Brassica napus* L., microprojected with the neomycin-phosphotransferase II (*nptII*) gene (Neuhaus et al. 1987).

A high-frequency regeneration and transformation between 27 and 51% determined by DNA dot blot analysis of primary haploid regenerants was reported. Pollen gene transfer was also tested, where pollen tubes were used as vectors with plasmids containing the coding regions of the *nptII* gene which were also transferred into several varieties of wheat and soybean (Shou et al. 2002; Ali et al. 2015). Other *in-planta*-based genetic transformation methods earlier included the use of germinated seeds as targeted plant materials for gene transfers, together with the use of wounded inflorescence and vacuum infiltration of floral plant organs such as stigmas for plant transformation (Desfeux et al. 2000; Li et al. 2017).

7.2.6 Mutagenesis

Genetic mutations are also one of the ways to induce dramatic changes in the metabolism, phenotype and physiological processes of many monocot and dicot plants. As new plant genetic variations are needed, increased diversity within species populations and newly modified traits which can confer built-in or inherent resistance to various stress factors are highly required. Nevertheless, to improve the growth and yield of soybeans, it is therefore important to understand the responses and developments of these plants when they are treated with chemical mutagens. Chromosome multiplication is usually applied and achieved through the use of chemicals such as colchicine, ethyl-methanesulfonate, oryzalin and amiprophos methyl (Feig et al. 1994; Hedebe et al. 2017; Wani 2017). There are limited studies done on the influence of these exogenous chemical mutagens on the growth, physiological and biochemical characteristics particularly to gain comprehensive insights on the growth attributes of the mutant plants. Supposedly, this technique could be more preferable due to the rising scepticism surrounding the production and use of genetically modified products developed by genetic engineering technology.

It has been widely proven that chemical mutagens induce gene mutation and physiological changes which can be detected and measured from seed germination, survival rates, plant morphology and fertility reductions or enhancements in first and further generation populations. Chemical mutagens are highly toxic to plants; hence high dose usage subsequently causes malformation or abnormalities in plants derived from seeds treated with the mutagen (Feig et al. 1994). According to Roychowdhury and Tah (2011), the higher the concentration of the chemical used the higher the biological damage or the higher the frequency of polyploidisation. These are the dynamics of a catch-22 situation that is inherent to the application of chemically induced mutagenesis. However, the use of chemical mutagenesis across the agricultural breeding fields by scientists/breeders is mainly intended to produce mutant plants that could serve as valuable plant breeding materials, and incorporating these mutant plants as stocks in other breeding systems to develop new and improved crop varieties (Jabeen and Mirza 2004).

7.2.7 Marker-Assisted Selection

Marker-assisted selection (MAS) is an indirect method of selecting traits of interest based on the marker linked to the traits rather than the traits themselves. RNA/DNA, biochemical and morphological markers linked to grain productivity, grain quality, biotic stress resistance or abiotic stress tolerance are selected to improve the efficiency in plant breeding through the precise transfer of genomic regions of interest (Das et al. 2017). However, genomic-wide selection has been widely tested to efficiently accelerate genetic improvement through direct selection during early juvenile plant stages. Performing predictions and selections early during plant growth was reported to increase the efficiency of potassium and nitrogen absorption under low soil nutrient availability in tropical maize (Fritsche-Neto et al. 2012). In rice, DNA-based molecular markers that are closely linked to resistance genes against biotic and abiotic stress were transferred to single genotypes (Das et al. 2017).

According to this report, the method can be used to rapidly accelerate the advancement of resistant cultivars, with the lowest number of generation breeding cycles precisely through the process of gene pyramiding, in addition to marker-assisted backcrossing, early generation selection and combined MAS. Nevertheless, sequence-based markers such as single nucleotide polymorphisms (SNP) and polymerase chain reaction (PCR)-based simple sequence repeat (SSR) or microsatellite markers remain the most preferred marker systems for genetic/genomic studies and crop breeding. Choudhary et al. (2012) reviewed about 2000 genomic SSR markers but with 30% polymorphic marker frequency as compared to other genera of this tribe. Based on these reports, SSRs exhibit polymorphism in terms of variation in the number of repeat units as revealed by the amplification of unique sequences flanking these repeat units, with their co-dominance inheritance suitable for genotyping segregating populations.

7.2.8 Qualitative Trait Loci

The identification of DNA markers, genes, and quantitative trait loci (QTLs) associated with particular traits is accomplished through QTL mapping (Das et al. 2017). QTL mapping enables a better understanding of the genetic linkages, control and inheritance of traits. This provides insights through the construction of linkage maps that enables breeders to detect specific chromosome parts that have candidate gene segments which can be used for stress resistance. This could also be used for choosing the best selective breeding strategy composed of genetic markers for a specific population. A large number of quantitative trait loci (QTLs) mapping studies for diverse crops species continue to provide an abundance of DNA marker–trait associations (Collard and Mackill 2008). QTL mapping signifies the basis of development of molecular markers for MAS regardless of the tool's inaccuracies, such as the replication levels of phenotypic information, population sizes and type and environmental effects as well as genotyping errors (Riday 2007).

The use of DNA markers for the identification of QTLs was a breakthrough in the characterisation of quantitative traits. Until recently, a wide range of QTLs governing various biotic and abiotic stress factors were identified at the genetic level in segregating mapping populations. Chen et al. (2008) mapped QTLs for heat stress tolerance using second filial generation plants (F_2), recombinant inbred line (RIL) and backcross inbred line (BIL) populations. Additionally, phenotypic variation ranging between

6.27 and 21.29% of the five identified QTLs found on chromosomes 5 and 9 was then reported, which conferred heat stress tolerance in rice, where chromosome 5 was even narrowed down from 23 Mbp to 331 Kbp. Yang et al. (2019) also reported a broad-sense heritability (h^2_b) for biological nitrogen fixation (BNF) ranging between 0.48 and 0.87% based on the genotype. Moreover, two new QTLs for BNF traits, *qBNF-16* and *qBNF-17*, were also identified, following the evaluation of soybean parent lines (with contrasting BNF traits) and 168 $F_{9:11}$ RILs under varied field conditions.

7.2.9 CRISPR-Cas9

Genome editing is a type of genetic engineering in which a DNA molecule is modified by the deletion or insertion of new specific fragments in the genome of a living organism. Gene editing serves as another new and highly innovative tool that gives scientists the ability to change the genetic makeup of organisms. Amongst some of the newly developed technologies, this RNA-guided clustered regularly interspaced short palindromic repeats (CRISPR) associated protein 9 (Cas9) system quickly rose to prominence due to its high precision of DNA alterations. The CRISPR-Cas9 gene editing technology is derived from the CRISPR-Cas adaptive immune system modules present in Archaebacteria and Eubacteria encoded by rapidly evolving and diverse operons (Makarova et al. 2011). Other genome editing technologies include zinc-finger nucleases (ZFNs) and transcription activator-like effectors and nucleases (TALENs) which emerged recently. ZFNs and TALENs use a strategy of tethering endonuclease catalytic domains to modular DNA binding proteins to induce specific genomic loci cleavage (Ran et al. 2013).

The use of these targeted nucleases, like in the case of CRISPR-Cas9, serves as one of the most powerful techniques that precisely mediate genomic alterations in plants. Several protocols are available that provide practical guidelines for identifying target sites, evaluate the cleaving process and analyse the occurrence of undesirable off-target cleavage activities. Depending on minimal off-target activity, genetic modifications could be rapidly achieved and cloned cell line regenerated to produce new plant varieties. In soybean, a few reports are available that tested and used the new CRISPR-Cas9-based genome editing. In one of the reports, this CRISPR-Cas9 technology was used to test both single and multiplexing targets in hairy roots, composite and tissue culture-based plant regeneration (Liu et al. 2019). So far, about two legume species, namely soybean and *Medicago truncatula*, were successfully mutated and genetically modified using Cas9 enzyme whose bioinformation was deposited in the online web tool for the fast identification of CRISPR-Cas9 target loci within gene models and genetic DNA sequences (Michno et al. 2015). As evidenced in other gene transfer tools, this technique may as well facilitate genomic alterations in soybean, various other legumes and many crop plant species in general.

7.3 OTHER UNCOMMON METHODS

There are several other transformation methods that were recently developed for use in crop improvement. These unique and rarely used genomic methods demonstrate the increasing interest of researchers in the use of chemicals. Such chemical methods are frequently used for gene transfers through *in vitro* or *in vivo* culture with the inclusion of calcium phosphates, diethyl-amino-ethyl-dextran (DEAE-D) and cationic polymers such as polyethyleneglycol (PEG). The applications of these chemical methods for gene delivery in plant host genomes are briefly discussed below.

7.3.1 Transfection via Calcium Phosphate Precipitation

Graham and van der Ed first reported the viral DNA transfection protocol using isotonic saline containing phosphate at low concentration mixed with calcium chloride to form calcium phosphate precipitation (Graham and van der Ed 1973). The DNA was co-precipitated with calcium phosphate making it easily absorbable by the human KB cells. This technique was since adopted for genetic alteration of both prokaryotic and eukaryotic cells. Nevertheless, this adoption took place after its modifications done by Orrantia and Chang (1990) showing that transient and stable expression of exogenous DNA was improved by internalised active endocytosis. In this pathway, DNA molecules to be transferred into new host cells were internalised through calcium phosphate precipitation and directly transported by intermediary vesicles derived from endocytic-lysosomal compartments and taken into the nucleus (Jin et al. 2014).

Calcium phosphate precipitation was developed when most of the assays were based on the DEAE-dextran method which only promoted the transformation of DNA from certain sources, especially simian virus 40 (SV40) (Orrantia and Chang 1990). Since then, this method has proven to be an easy-to-use, safe and a highly cost-effective tool for gene transfer and expression in many types of living cells. This remained the case despite challenges such as reaction time, temperature and the physical and chemical conditions that seriously influence the formation of DNA-hydroxyapatite calcium/phosphate particles (Jin et al. 2014). DNA absorption and protection against the attack of nucleases associated with hydroxyapatite particles appears to be one of the main reasons why this technique has not been widely used for the transformation of plant species. Furthermore, the technique is prone to insufficient efficiencies and very poor reproducibility compared to other gene transfer methods (Sokolova et al. 2006).

7.3.2 Diethyl Amino Ethyl (DEAE) Dextran-Mediated DNA Transfer

The DEAE-dextran gene transfer is also one of the techniques that is widely used for the transformation of mammalian cells. DEAE-dextran is a polycationic derivative of dextran, mixed with diethylaminoethyl groups which are positively charged molecules. This mixture of chemical compounds was found to easily access and penetrate through negatively charged membrane surfaces of cells. This means that DEAE-dextran that are successfully complexed with exogenous DNA molecules can be easily imported through the cells by endocytosis (Ebbesen 1974). This approach was reported to be the first chemical vector for exogenous DNA delivery in 1965 by Veheri and Pagano (Holter et al. 1989). In one of the earliest reports, DEAE-dextran at 0.1 and 1 μg/mL was used for the binding and up taking of exogenous DNA by nuclei isolated from soybean cultivar SB-1 grown on Gamborg's B_5 medium (Ohyama 1978). This chemical was then later reported to play a role in the genetic transformation of cell-walled plant and algae cells for DNA delivery (Ortiz-Matamoros et al. 2017). DEAE-dextran-based genetic transformation appears to yield higher transformation efficiencies in specific mammalian cells than in plant cells and tissues. Furthermore, there is clear evidence on its applications that it could contribute by establishing a gene transferring system into mammalian cells rather than in plant cells, given a large number of reports related to mammalian cell transformation.

7.3.3 Polyethylene Glycol (PEG)-Mediated DNA Transfer

The genetic transformation of plant cells using polyethylene glycol (PEG) as an osmotic, physiologic and molecular influencer has been widely reported compared to the DEAE-dextran-mediated gene transfer discussed above. PEG is an inert amphophilic polymer consisting of repeating subunits of ethylene glycol prepared as a copolymer useful for biomedical and tissue engineering applications (Chen et al. 2005). This compound can be utilised during transformation for improving protoplast cell aggregation, fusion and protein condensation through the effects of depletion, destabilisation and bilayer dehydration (Ross and Hui 1999). It was, furthermore, indicated that higher concentrations of PEG in a cell culture lead to membrane lipid bilayer disruptions due to excessive dehydration; meanwhile, lower amounts result in membrane fusion between adjacent cells (Yang et al. 1997). Even though PEG is rarely used in soybean transformation, a number of reports demonstrated its role as an osmoticum during somatic embryogenesis.

Higher matured embryo quantities were obtained from cotyledons of immature seeds in soybean cultivar Bragg and IAS5 on MSB_5 medium supplemented with 25 g/L PEG-8000 (Korbes and Droste 2005). In a similar study, variations were observed on the quantity of mature embryos recovered, their germination rates (shoot and root emergence) and conversion into plants when 5% PEG was supplemented in a liquid Finer and Nagasawa medium-based histodifferentiation/maturation medium (Walker and Parrott 2001). In most cases, a protocol for PEG-mediated genetic transformation involves the use of protoplasts in *in vitro* cultures. The PEG-mediated transformation of protoplasts has been the most common approach used for genetic improvement of monocot and dicot plants. But very few reports are available on legume crops, such as soybeans (Klebe et al. 1983; Hayashimoto et al. 1990; Kofer et al. 1998).

7.4 STATUS OF *A. RHIZOGENES* IN PLANT TRANSFORMATION

Agrobacterium tumefaciens has long occupied a superior position in contrast to other biological natural vectors for the genetic transformation of many monocots and dicots. *A. rhizogenes* also serves as one of the most critically important and occasionally used Gram-negative bacteria for the genetic engineering of dicotyledonous plants. However, like *A. tumefaciens*, *A. rhizogenes* is less culture and labour intensive. *A. rhizogenes*' utilisation in modern breeding technology remains limited to some cultivars and species due to its transgenic plant regeneration and recovery challenges. Genetic transformation with *A. rhizogenes* involves recovering transgenic plantlets from transformed hairy roots after the transfer and integration of a piece of its root inducing (Ri) plasmid DNA containing the root locus (*rol*) genes. These *rol* genes are responsible for causing hairy root diseases in host plants (Bevan and Chilton 1982).

A. rhizogenes is still marked by the controversies surrounding its classification as to whether it is recognised to belong to the genus *Rhizobium* or *Agrobacterium* (Young eta a. 2001) as previously discussed in Chapter 3. The technique continues to be gradually used to shuttle foreign genes derived from different sources (animals, bacteria, fungi, plants and viruses) into hosts to produce new plant lines. It became more popular and preferable for use in the manufacturing of pharmaceutical compounds, studying gene functions and root biology, than being used for the production of transgenic plants. Artemisinin and morphinan alkaloids, for instance, were respectively produced from *Artemisia annua* and *Papaver orientale* using root cultures following a successful *A. rhizogenes*-mediated genetic transformation procedure (Bandaranayake and Yoder 2018).

7.5 PROS AND CONS OF ARTIFICIAL GENE TRANSFER METHODS

The application of gene transfer procedures for the functional analysis of eukaryotic and prokaryotic genes has greatly facilitated our understanding of the mechanisms involved in gene regulation. Plant regulatory elements involved in transcriptional regulation have been analysed, especially in forage and grain crops. This was particularly done using *Agrobacterium tumefaciens*-based transformation systems, including the chemical and physical gene transfer methods. Studying these regulations of gene expressions in crops enabled the development of homologous gene transfer systems that have been used to circumvent the difficulties encountered during plant improvement, as well as counteracting growth- and yield-associated challenges. A number of those beneficial genetic improvements will be thoroughly discussed in the next chapters. However, the establishment of a highly efficient and reproducible procedure for crop propagation and fertile plant regeneration is still yet to be achieved, particularly in soybean.

Apart from the challenges faced due to gene expression and *in vitro* regeneration conditions as discussed in the previous chapter (Chapter 6), most of these chemical and physical transformation methods are faced with serious hurdles. Both biolistic/bombardment and PEG-mediated transformation techniques, for instance, can only accomplish their goals if plant cell wall barriers are removed to enable the penetration of exogenous DNA molecules to effect gene transfer and expression (Kofer et al. 1998). When protoplast cultures were first established, this became a brilliant solution to overcoming cell wall barriers in plants transformed with these tools. But later, the technique's inefficiencies in cell aggregation, fusion and difficulties in plant regeneration from transformed protoplasts emerged as new major challenges for the routine application of this technology. There are serious difficulties in making sure that large molecules containing desired sequences (DNA and RNA for example) are successfully passed through the lipid bilayer of a selectively permeable membrane enclosing cells and cellular compartments, making sure that these transformed cells are then regenerated into individual plants.

Electroporation, therefore, serves as one of the most simple, quick and highly efficient genetic transformation methods also used to bypass these barriers in a variety of grain and vegetable crops. Unfortunately, this technique can only be efficiently applied to protoplast cultures in order to effect genetic alterations. As indicated previously, protoplast cultures are prone to regeneration inefficiencies (Darbani et al. 2008). The demand for novel strategies to manipulate and incorporate specific genetic fragments into crops to improve growth and development characteristics in agreement with the needs of consumers and society at large is growing. Plant transformation, whether performed by biological, chemical or physical means, must be improved to effectively deal with the serious difficulties seen in their protocols. More research and protocol optimisations are still necessary. Scientists should, thus, continue to optimise existing protocols or develop new gene transfer methods for the proper implementation and routine applications of these technologies across all genotypes.

7.6 SUMMARY

Most conventional breeders employ backcross and multiple generations of ancestors (pedigree) to increase the certainty of pure lines and accelerate the expression of desirable traits in the progenies. These improvement efforts are continuously investigated and implemented due to the production constraints resulting in growth and yield inhibitions experienced by various crops, especially legumes like chickpea, cowpea and soybean. However, several biotechnological strategies have also been developed to efficiently increase crop growth development and productivity. The introgression of novel gene combinations using genetic engineering or recombinant DNA technology that allows for the use of specific transgenes normally transformed in binary plasmid vectors of bacterial origin or those aided by chemical and physical means are continuously tested. These technologies usually involve the identification, isolation, reconstruction and transfection of exogenous DNA segments into host plants carried out in step-wise laboratory procedures in the attempt to advance crops' nutritional qualities and tolerance to stress factors.

Genetic transformation holds great potential to be systematically applied and optimised for the development of new agronomically useful crop varieties. Attempts to produce new breeding materials in legumes using a variety of physical and chemical transformation techniques discussed above have been pursued. But most of these crops have proved to be highly recalcitrant or resistant to the currently employed genetic transformation techniques. This is so because the increase in legume yields still relies upon agricultural chemical applications (fertilisers, herbicides and pesticides) which generate various economic and ecological problems, particularly the problem of environmental pollution. A novel biolistic-mediated genetic improvement system in the legume cowpea was reported (Ivo et al. 2008). The system combined the use of herbicide imazapyr to select transgenic meristematic cells after the biolistic introgression of mutated *ahas* gene coding for acetohydroxyacid synthase under the control of *ahas* '5 regulatory sequence.

Biotic stress-resistant features adapted from several legumes and transformation systems were reproducibly used to produce transgenic cowpeas. These transgenic cowpeas contained the *bar* gene and expressed phosphinothricin acetyl transferase (PAT) enzyme activity that confers resistance towards phosphinothricin and bialaphos with the normal production of tropane and alkaloids (Davison and Ammann 2017). Additionally, an improved sonication-assisted *Agrobacterium*-mediated gene delivery using cotyledonary node explants derived from imbibed cowpea seeds was also reported (Bett et al. 2019). Now a much faster and easier DNA modifying technology is the clustered regularly interspaced short palindromic repeat (CRISPR) and flanked by CRISPR-associated genes (Cas9). CRISPR-Cas9 is a unique set of partially palindromic repeated DNA sequence found in the genome of bacteria and other microorganisms, emerged in recent years to induce targeted DNA double-stranded breaks at specific genomic loci.

Among these, the small RNAs guided Cas9 nucleases serve as a high-throughput, efficient and highly specific gene-editing tool for a variety of cell types and organisms using a Watson-Crick base pairing (Ran et al. 2013). It was demonstrated that it is possible to use a CRISPR-Cas9 system application in legume (e.g., cowpea or soybean) genome editing for mutant generation by disrupting the representative symbiotic nitrogen fixation (SNF) genes (Ji et al. 2019). Therefore, looking at all efforts described above, and the global challenges such as climate change, food insecurity, ever-increasing human world population size and malnutrition demands, rapid advances in agricultural technology are indeed necessary. Novel strategies to manipulate and improve crop growth and yield are a major prerequisite in solving these global challenges.

7.7 ABBREVIATIONS

BIL	Backcross inbred line
BNF	Biological nitrogen fixation
CRISPR-Cas9	Clustered regularly interspaced short palindromic repeats – associated protein9
DEAE-D	Diethyl amino ethyl-dextran
DNase	Deoxyribonuclease enzyme
dsDNA	Double stranded DNA
Gus	β-glucuronidase
LPTC	Liposome transfection complexes
MAS	Marker-assisted selection
nptII	Neomycin phosphotransferase II
PCR	Polymerase chain reaction
PEG	Polyethylene glycol
QTL	Quantitative trait loci
RNA	Ribonucleic acid

SNP	Single nucleotide polymorphism
SSR	Simple sequence repeat
SV40	Simian virus 40
TALENs	Transcription activator-like effectors and nucleases
ZFNs	Zinc-finger nucleases

REFERENCES

Ali A, Bang SW, Chung S-M and Staub JE. (2015). Plant transformation via pollen tube-mediated gene transfer. *Plant Molecular Biology Reporter* 3, 742–747.

Bandaranayake PCG and Yoder JI. (2018). Factors affecting the efficiency of *Rhizobium rhizogenes* root transformation of the root parasitic plant *Triphysaria versicolor* and its host *Arabidopsis thaliana*. *Plant Methods* 14(61), 1–9.

Behrooz D, Safar F, Mahmoud T, Saeed Z, Shahin N and Stewart Jr CN. (2008). DNA-delivery methods to produce transgenic plants. *Biotechnology* 7, 385–402.

Bett B, GOllasch S, Moore A, Harding R and Higgins TJV. (2019). An improved transformation system for cowpea (*Vigna unguiculata* L. Walp) via sonication and a kanamycin-geneticin selection regime. *Frontiers in Plant Science* 10(219), 1–10.

Bevan MW and Chilton MD. (1982). T-DNA of the *Agrobacterium* Ti and Ri- plasmid. *Annual Review of Genetics* 16, 357–384.

Boynton JE, Gillham NW, Harris EH, Hosler JP, Johnson AM, Jones AR, Randolph-Anderson BL, Robertson D, Klein TM, Shark KB and Sanford. (1988). Chloroplast transformation in *Chlamydomonas* with hig velocity microprojectiles. *Science* 240(4858), 1534–1538.

Chen J, Spear SK, Huddlestron JG and Rogers RD. (2005). Polyethylene glycol and solutions of polyethylene glycol as green reaction media. *Green Chemistry* 7(2), 64–82.

Chen Q, Yu S, Li C and Mou T. (2008). Identification of QTLs for heat tolerance at flowering stage in rice. *Zhongguo Nong Ye Ke Xue* 10(1), 315–321.

Choudhary AK, Gaur R, Gupta S and Bhatia S. (2012). EST-derived genic molecular marker: Development and utilization for generating an advanced transcript map of chickpea. *Theoretical and Applied Genetics* 124(8), 1449–1462.

Christou P. (1994). The biotechnology of crop legumes. *Euphytica* 74, 165–185.

Collard BCY and Mackill DJ. (2008). Marker-assisted selection: An approach for precision plant breeding in the twenty-first century. *Philosophical Transactions of the Royal Society B- Biological Science* 363(1491), 557–572.

Darbani B, Farajnia S, Toorchi M, Zakerbostanidad S, Noeparvar S and Sterwart Jr NC. (2008). DNA-delivery methods to produce transgenic plants. *Biotechnology* 7(3), 385–402.

Das G, Patra JK and Baek KH. (2017). A molecular tool for the development of stress resistant and quality of rice through gene stacking. *Frontiers in Plant Science* 8(985), 1–9.

Davison J and Ammann K. (2017). New GMO regulations for old: Determining a new future for EU crop biotechnology. *GM Crops Food* 8(1), 13–34.

Desfeux C, Clough SJ and Bent AF. (2000). Female reproductive tissues are the primary target of *Agrobacterium*-mediated transformation by the *Arabidopsis* floral dip method. *Plant Physiology* 123(3), 895–904.

Deshayes A, Herrera-Estrella L and Caboche M. (1985). Liposome-mediated transformation of tabacco mesophyll protoplasts by an *Escherichia coli* plasmid. *The Embo Journal* 4(11), 2731–2737.

Doukas AG and Kollias N. (2004). Transdermal drug delivery with a pressure wave. *Advanced Drug Delivery Review* 56, 559–579.

Ebbesen P. (1974). Influence of Deae-dextran, polybrene, dextran and dextran sulphate on spontaneous leukaemia development in AKR mice and virus induced leukaemia in BALB/c mice. *British Journal of Cancer* 30, 68–72.

Feig DI, Somers LC and Loeb LA. (1994). Reverse chemical mutagenesis: Identification of the mutagenic lessions resulting from reactive oxygen species-mediated damage to DNA. *Proceedings of the National Academy of Sciences of the United States of America* 91(14), 6609–6613.

Fritsche-Neto R, Ferreira LR, Ferreira FA, da Silva AA and Do Vale JC. (2012). Breeding for weed management. In Fritsche-Neto R and Borem A (eds), *Plant Breeding for Biotic Stress Resistance*. Springer-Verlag, Berlin, Heidelberg. pp. 137–164.

Graham FL and van der Eb AB. (1973). A new technique for the assay of infectivity to human adenovirus 5 DNA. *Virology* 52(2), 456–467.

Hadebe ST, Modi AT and Shimelis HA. (2017). Determination of optimum ethylmethanesulfonate conditions for chemical mutagenesis of selected vernonia (Centrapalus pauciflorus) accessions. *South African Journal of Plant and Soil* 34(4), 311–317.

Hassan S, Ahmad A, Batool F, Rashid and Husnain T. (2021). Genetic modification of *Gossypium arboreum* universal stress protein (GUSP1) improves drought tolerance in transgenic cotton (*Gossypium hirsutum*). *Physiology and Molocular Biolology of Plants* 27, 1779–1794.

Hayashimoto A, Li Z and Murai M. (1990). A polyethylene glycol-mediated protoplast transformation system for production of fertile transgenic rice plants. *Plant Physiology* 93, 857–863.

Holter W, Fordis CM and Howard BH. (1989). Efficient gene transfer by sequential treatment of mammalian cells with DEAE-dextran and deoxyribonucleic acid. *Experimental Cell Research* 184, 546–551.

Ivo NL, Nascimento CP, Vieira LS, Campos FAP and Aragao FJL. (2008). Biolistic-mediated genetic transformation of cowpea (*Vigna unguiculata*) and stable Mendelian inheritance of transgenes. *Plant Cell Reports* 27(9), 1475–1483.

Jabeen N and Mirza B. (2004). Ethyl methane sulfonate induces morphological mutations in *Capsicum annum*. *International Journal of Agriculture and Biology* 6(2), 340–345.

Ji J, Zhang C, Sun Z, Wang L, Daunmu D and Fan Q. (2019). Genome editing in cowpea Vigna unguiculata using CRISPR-Cas9. *International Journal of Molecular Sciences* 20(247), 1–13.

Jin L, Zeng X, Liu M, Deng Y and He N. (2014). Current progress in gene delivery technology based on chemical methods and nanocarriers. *Theranostics* 4(3), 240–255.

Kaeppler HF, Gu W, Somers DA, Rines HW and Cockburn AF. (1990). Silicon carbide fiber-mediated DNA delivery into plant cells. *Plant Cell Reports* 9, 415–418.

Killing N, Gonzalez-Blanco P, Gowda S, Martini X and Etxeberria E. (2021). Plant functional genomics in a few days: Laser-assisted delivery of double-stranded RNA to higher plants. *Plants (Basel)* 10(93), 1–8.

Klebe RJ, Harriss JV, Sharp ZD and Douglas MG. (1983). A general method for polyethylene-glycol induced genetic transformation of bacteria and yeast. *Gene* 25(2–3), 333–341.

Ko TS, Nelson RL and Korban S. (2004). Screening of multiple soybean cultivars (MG00 to MGVIII) for somatic embryogenesis following *Agrobacterium*-mediated transformation of immature cotyledons. *Crop Science* 44, 1825–1831.

Kofer W, Eibl C, Steinmuller K and Koop H-V. (1998). PEG-mediated plastid transformation in higher plants. *In Vitro Cell and Developmental Biology- Plant* 34, 303–309.

Korbes AP and Droste A. (2005). Carbon sources and polyethylene glycol on soybean somatic embryo conversion. *Pesquisa Agropecuaria Brasileira* 40(3), 211–216.

Li S, Cong Y, Liu Y, Wang T, Shuai Q, Chen N, Gai J and Li Y. (2017). Optimization of *Agrobacterium*-mediated transformation in soybean. *Frontiers in Plant Science* 8(246), 1–15.

Lin YC, Chen L-Y, Chen C-H, Liu Y-K, Hsu W-T, Ho L-P and Liao K-W. (2012). A soybean oil-based liposome-polymer transfection complex as a delivery system for DNA and subunit vaccines. *Journal of Nanomaterials* Article ID 427306, 1–12.

Liu J, Gunapati S, Mihelich NT, Stec AO, Michno J-M and Stupar RM. (2019). Genome editing in soybean with CRISPR/Cas9. *Methods in Molecular Biology* 1917, 217–234.

Makarova KS, Haft DH, Barrangou R, Brouns SJJ, Charpentier E, Horvath P, Moineau S, Mojica FJM, Wolf YI, Yakunin AF, van der Oust J and Koonin EV. (2011). Evolution and classification of the CRISPR-Cas systems. *Nature Reviews Microbiology* 9, 467–477.

Meacham JM, Durvasula K, Degertekin FL and Fedorov AG. (2014). Physical methods for intracellular delvery: Practical aspects from laboratory use to industrial-scale processing. *Journal of Laboratory Automation* 19(1), 1–18.

Michno J-M, Wang X, Liu J, Curtin SJ, Kono TJY and Stupar RM. (2015). CRISPR/Cas mutagenesis of soybean and *Medicago truncatula* using a new web-tool and a modified Cas9 enzyme. *Biotechnology in Agriculture and the Food Chain* 6(4), 243–252.

Neuhaus G, Spangenberg G, Mittelsten Scheid O and Schweiger H. G. (1987). Transgenic rapeseed plants obtained by the microinjection of DNA into microspore-derived embryoids. *Theoretical and Applied Genetics* 75, 30–36.

Ohyama K. (1978). DNA binding and uptake by nuclei isolated from plant protoplasts. *Plant Physiology* 61, 515–520.

Olhoft PM and Somers DA. (2001). L-cysteine increase *Agrobacterium*-mediated T-DNA delivery into soybean cotyledonary-node cells. *Plant Cell Reports* 20, 706–711.

Orrantia E and Chang PL. (1990). Intracellular distribution of DNA internalized through calcium phosphate precipitation. *Experimental Cell Research* 190(2), 170–174.

Ortiz-Matamoros MF, Villanvera MA and Islas-Flores T. (2017). Genetic transformation of cell-walled plant and algae cells: Delivering DNA through the cell wall. *Briefings in Functional Genomics* 17(1), 26–33.

Ran FA, Hsu PD, Wright J, Agarwala V, Scott DA and Zheng F. (2013). Genome engineering using the CRISPR-Cas9 system. *Nature Protocols* 8, 2281–2308.

Rech EL, Vianna RG and Aragao FJL. (2008). High efficiency transformation by biolistics of soybean, common bean and cotton transgenic plants. *Nature Protocols* 3(3), 410–418.

Riday H. (2007). Marker-assisted selection in legumes. *Lotus News Letters* 37(3), 102.

Rivera AL, Gomez-Lim M, Fernandez F and Loske AM. (2012). Physical methods for genetic plant transformation. *Physics of Life Review* 9, 308–345.

Ross PC and Hui SW. (1999). Polyethylene glycol enhances lipoplex-cell association and lipofection. *Biochimica et Biophysica Acta (BBA): Biomembranes* 1421(2), 273–283.

Roychowdhury R and Tah J. (2011). Chemical mutagen action on seed germination and related agro-metrical traits in M_1 *Dianthus* generation. *Current Botany* 2(8), 19–23.

Serik O, Ainur I, Murat K, Tetsuo M and Masaki I. (1996). Silicon carbide fiber-mediated DNA delivery into cells of wheat (*Triticum aestivum* L.) mature embryos. *Plant Cell Reports* 16, 133–136.

Shiokawa D and Tanuma S. (2001). Characterization of human DNase 1 family endonucleases and activation of DNase gamma during apoptosis. *Biochemistry* 40(1), 143–152.

Shou H, Palmer RG and Wang K. (2002). Irreproducibility of the soybean pollen-tube pathway transformation procedure. *Plant Molecular Biology Reporter* 20, 325–334.

Sokolova V, Radtke I, Heuman R and Epple M. (2006). Effective transfection of cells with multi-shell calcium phosphate-DNA particles. *Biomaterials* 27(16), 3147–3153.

Walker DR and Parrott WA. (2001). Effect of polyethylene glycol and sugar alcohols on soybean somatic embryo germination and conversion. *Plant Cell, Tissue and Organ Culture* 64, 55–62.

Wani MR. (2017). Induced chlorophyll mutations, comparative mutagen effectivieness and efficiency of chemical mutagens in lentils (*Lens culinaris* Medik). *Asian Journal of Plant Sciences* 16(4), 221–226.

Yan B, Reddy MSS, Collins CB and Dinkins RD. (2000). *Agrobacterium tumefaciens*-mediated transformation of soybean [*Glycine max* (L.) Merrill] using immature zygotic cotyledon explants. *Plant Cell Reports* 19, 1090–1097.

Yang Q, Guo Y, Li L and Hui SW. (1997). Effects of lipid headgroup and packaging stress on polyethylene glycol-induced phospholipid vesicle aggregation and fusion. *Biophysical Journal* 73, 277–282.

Yang Q, Yang Y, Xu R, Lv H and Liao H. (2019). Genetic analysis and mapping QTLs for soybean biological nitrogen fixation traits under varied field conditions. *Frontier in Plant Science* 10(75), 1–11.

Young JM, Kuykendall LD, Martinez-Romero E, Kerr A and Sawada H. (2001). A revision of *Rhizobium* Frank 1889, with an emended description of the genus, and the inclusion of all species of *Agrobacterium* Conn 1942 and *Allorhizobium undicola* de Lajudie et al. 1998 as new combinations: *Rhizobium radiobacter, R-rhizogenes, R-rubi, R-undicola and R-vitis. International Journal of Systematic and Evolutionary Microbiology* 51, 89–103.

Zeng P, Vadnais DA, Zhang Z and Polacco JC. (2004). Refined glufosinate selection in *Agrobacterium*-mediated transformation of soybean [*Glycine max* (L.) Merrill]. *Plant Cell Reports* 22(7), 478–482.

8 Molecular Aspects of Indirect Gene Transfer via *A. tumefaciens*

8.1 INTRODUCTION

The use of *Agrobacterium tumefaciens* in the genetic improvement of plants is one of the most common and widely available pathways for vector-mediated genetic transformation. To follow operational sequence, an *Agrobacterium* strain is identified, its genome is cloned with a *cis* or binary plasmid vector construct containing desirable genes, it is utilised to infect target tissues of host plant and then it incorporates and expresses its T-DNA into cells of the host plant. The whole induction and genetic engineering of both the bacterium and plants have been reviewed, especially from the developmental biology perspective. As the biology of the *Agrobacterium* and conditions of *in vitro* transformation were described in Chapters 3 and 4, this chapter will only focus on the molecular aspects of *Agrobacterium*-mediated genetic transformation in plants, with more reference to soybean. Genetic transformation using *A. tumefaciens* is very unique in that this soil plant pathogen naturally infects plants through their wound site to deliver its T-DNA. *Agrobacterium* uses the bacterial type IV secretion system (T4SS) which is a complex macromolecular nanomachine that translocates its genetic materials and effector proteins from the bacterium's cytoplasm into targeted plant cells (Kahrstrom 2014).

The system constitutes various tumourigenic genes and other promotors to transformatively interact with host cells. At the moment, oncogenic sequences in T-DNA regions of the binary vectors can be replaced with genes of interest due to advancements and innovation in molecular biology (Hwang et al. 2017). *Agrobacterium*-mediated transformation includes the following main operational processes: bacterial cell rejuvenation, explant preparation and infection, co-cultivation, shoot induction and the selection and identification of transformed plantlets for further growth and development. The majority of these stages, if not all, require gene activation and expression almost at all levels in order to achieve cell development programmes required for growing progenies/regenerants containing the new gene of interest. Therefore, genetic and molecular investigations have been done in the past decades to identify all of the critical molecular aspects involved in the various stages of genetic transformation using *A. tumefaciens*. Furthermore, these molecular analyses have been done also in recalcitrant legume crops such as soybean and other agriculturally important plants.

DOI: 10.1201/b22829-8

8.2 MOLECULAR MECHANISM OF *AGROBACTERIUM*-MEDIATED GENE TRANSFER

More than three decades have elapsed since the introduction of genetically modified plants established through the use of this technology. Some of the first successful transformation recoveries included the production of transgenic soybean plants that showed tolerance to the herbicide glyphosate using pTiT37-SE with a pMON9749 plasmid vector (Hinchee et al. 1988), followed by the production of a large number of transgenic monocotyledonous varieties in species such as rice, wheat and maize using other superior strains and vectors (Sood et al. 2011). Numerous studies have clearly showed that it is now more possible to transform monocot plants than dicots species using a suitable *Agrobacterium* strain, proper culture conditions, tailor-made promoters and gene constructs. Proving that, this technology is one of the most well-established methods used to produce a variety of transgenic products derived from both monocots and dicots used to cater for human and animal needs. Since plant transformation has grown in popularity, several biotechnological inventions have emerged ranging from the generation of recombinant proteins to the extraction of useful secondary compounds.

These applications used transgenic plants as bioreactors. The advancements were accomplished with the help of plant tissue culture, especially through cell suspension and callus culture. Furthermore, developments relating to soybean transformation or the improvement of related recalcitrant legumes, on the other hand, had not been well catered for, particularly through this biotechnological tool. For example, more breakthroughs by global researchers were made in studies directed towards the genetic improvement of cereal grain crops as sources of food and oil, and conferring tolerance to drought and biotic stress constraints. The margins of transformation efficiencies in these monocots are not comparable to the efficiency of production in transgenic soybean plants. Genetic transformation protocols that are based on *A. tumefaciens* or what is now known as vector-mediated methods continue to be explored for the trait improvement of many crops in the enhancements of nutritional content and resistance to environmental factors and diseases.

To date, the technique has succeeded in the generation of high yielding transgenic cultivars, particularly for corn, rice and some of the legume grains such as chickpea, cowpea and numerous regional cultivars of soybean (Ko et al. 2004; Mehrotra et al. 2011; Patel et al. 2013). The use of molecular components involved in plant transformation is considered an essential constituent of crop improvement, from basic studies to the development of novel commercial varieties that contain new combinations of genes for improved yield, nutritional value and stress tolerance. As a result, it is now considered the most economic and highly efficient method of plant breeding in modern genetics and molecular biology.

8.2.1 PATHOGENESIS

Gene transfer from the soil bacterium to plants occurs naturally using the tumour-inducing (Ti) plasmid present in *A. tumefaciens* or even the hairy root-inducing (Ri) plasmid found in *A. rhizogenes*. This pathogenic character of *Agrobacterium* is used during plant transformation to transfer Ti/Ri-plasmids and integrate them into the plant nuclear genome. This natural phytopathogenic ability was harnessed by scientists decades ago and is still used today for the purpose of plant genetic engineering. However, the incorporation of T-DNA into host plant cells requires genetic determinants of both the bacterium and plant (Gelvin 2003). In monocotyledonous plants, transformation initially failed to take place due to severe inefficiencies that gave the impression that these plants did not form part of its host-range system in causing a crown gall disease. However, key modifications in culture conditions, strains of *A. tumefaciens* used and the application of highly proliferative tissue organs required as explants to induce a wound response for infection later made transformation in these species possible (Hinchee et al. 1988; Zhang et al. 1999).

As is well-known, this process takes place highly efficiently in monocotyledonous plants compared to in dicotyledonous species. Therefore, strains successfully used for the genetic transformation of monocot plants are the same as those previously and currently employed for the transformation of dicotyledons. This shows that this gene transfer method still continues to take advantage of the presence of the Ti/Ri-plasmids found in *A. tumefaciens* and *A. rhizogenes* to transfer T-DNA segments of their vector across different plant hosts. The bacterium is capable of transferring this portion of its genome into plant hosts upon infection on the wounded plant cells. T-DNA is translocated into the plant cell nucleus and becomes stably integrated into the chromosomes (Lee and Gelvin 2017). This process is efficiently mediated collectively by the virulence (*vir*) genes (*virA*, *virB*, *virD*, *virG*, *virC* and *virE*) illustrated in Figure 8.1.

8.2.2 T-DNA

Agrobacterium tumefaciens-mediated plant transformation is a highly complex and evolving process that relies mainly upon successful T-DNA transfer and integration into host plant cells. The larger tumour-inducing and hairy root-inducing plasmid resident in *Agrobacterium* contain what is known as the T-regions of approximately 10 to 30 kilo base-pairs (kbp) in size (Gelvin 2003). This region is where the T-DNA is located, making up about 10% of the Ti/Ri-plasmids. Barker et al. (1983) reported that some Ti-plasmids contain a single T-region; meanwhile others may have multiple T-regions possessing the T-DNA processing and export molecular machinery. The protein products encoded by these operons that are found in the *vir* region of the vectors function to increase the T-DNA transfer efficacy into the host cells.

During this transfer, the T-DNA also serves as the only mobile segment of the Ti-plasmid that is translocated and integrated into the host plant's genome during pathogenesis (Gelvin 2003). This transfer process is regulated by the activity of some *vir* genes carried by the vector plasmid as illustrated in Figure 8.1. Furthermore, the T-DNA region of the plasmid also contains the oncogenic and opine genes that are responsible for encoding enzymes involved in the synthesis of hormones (auxins and cytokinins) playing a critical role in tumour induction and the formation of novel amino acids-sugar conjugates used by the bacteria as a source of food (McCullen and Binns 2006). Further details on this process were described in Chapter 3.

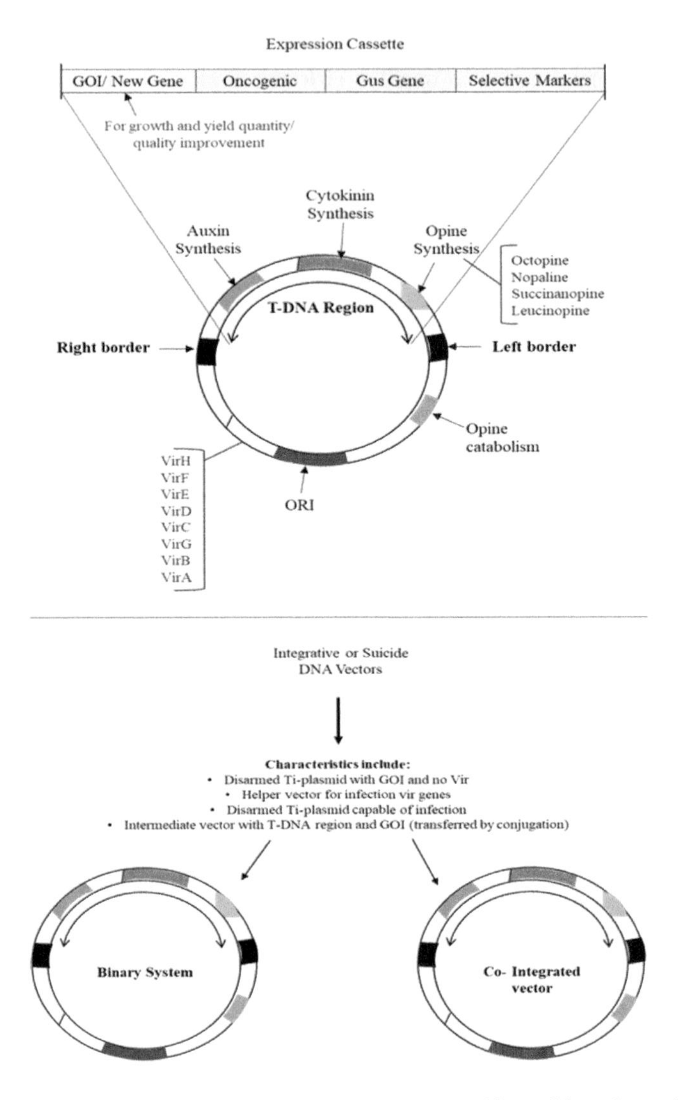

FIGURE 8.1 Examples of binary/helper vector systems used for stable and transient genetic transformation of monocotyledonous and dicotyledonous plants.

These macromolecules play an important role in mediating the transfer of DNA molecule and protein complexes between the bacterial and plant cell membrane, in conjunction with the *vir* gene components of the Ti-plasmid. According to the above-mentioned report by McCullen and Binns, these components of host recognition and macromolecular transfer of virulence-mediated effectors represent a very critical stage in the successful transformation of plant cells by *A. tumefaciens* (for example, octopine, nolapine, etc.) as previously discussed. T-DNA regions contain genes used for opine synthesis (octopine, nolapine, etc.) that are produced according to the strain of *Agrobacterium*. Typically, different strains synthesise and catabolise a specific set of opines. Moreover, genes of interests (*goi*), selection markers and other expression cassettes required for efficient plant transformation are also found in this DNA region (Figure 8.1).

8.2.3 Gene of Interest

The concept of 'tumour-inducing principle', that bacterial DNA molecules can be stably transferred and expressed in plant genomes, led to further elucidations of the molecular mechanisms of *Agrobacterium*-mediated genetic transformation in plants (Gelvin 2017). Scientists then continued to identify and clone different segments of the DNA into the T-regions of the Ri/Ti-plasmids for incorporation in host genomes. More specifically, the genes of interest can be cloned into the T-DNA region of the plasmid and transferred to plant cells where they are integrated in plants' nuclear DNA. This molecular evidence led to the general use of both Ri- and Ti-plasmid vectors to introduce foreign genes into plant cells. However, the cloning of genes into the Ti-plasmid requires that this is done in *cis* elements with virulence genes or cloned into a separate replicon as a binary vector without *vir* genes (Lee and Gelvin 2017). To achieve this, usually a broad host range plasmid that self-transfers, replicates and stably integrates in the host is required.

Broad host range plasmids are able to replicate in *Escherichia coli* where they are initially cloned and then inserted in *Agrobacterium* cells for the purpose of carrying exogenous genes of interest, together with selectable marker genes (Jain and Srivastava 2013). The gene of interest is flanked by homologous regulatory sequences found on the plasmid. Specific *goi* can be isolated and replicated to produce a large number of copies. These foreign genes can also be genetically altered before being transferred into host plant cells. Any genetic transformation procedure will be considered successful if it ensures that the *goi* is efficiently transferred and expressed, remaining functional as well as heritable in all host plants containing it. *Agrobacterium*-mediated genetic transformation was used to alleviate the imbalance effects in amino acid composition that restrict the enhanced biological value of legumes. Foreign genes of interest such as the Brazil nut 2S albumin (BNA) and its homologous sunflower genes were transferred into grain legumes such as narbon bean (*Vicia narbonensis* L.), lupins (*Lupinus angustifolius* L.) and soybean (*Glycine max* L. Merr.).

The approach was intended to increase free methionine (Met), increase Met containing endogenous proteins and transfer *goi* encoding Met-rich plant proteins (Muntz et al. 1998). Expression of the barley (*Hordeum vulgare*) abundant protein (*HVA1*) gene regulated by the rice *actin1* gene promoter was also confirmed in common bean (*Phaseolus vulgare* L.). This gene conferred tolerance to drought stress with a corresponding increase in root length of transgenic plants (Kwapata et al. 2012). Plant regeneration through somatic embryogenesis coupled with *A. tumefaciens*-mediated gene delivery was achieved through the preculture of embryo axes in common bean. Identified transformed cells contained the *bialaphos resistance* (*bar*) gene under the constitutive 35S promoter used as a selectable marker (Song et al. 2020). These successful studies on enriched Met-synthesis, *HVA1* drought tolerance and *bar* genes document that abiotic and biotic stress-associated problems in agriculture can be resolved by incorporating the various genes of interest through the use of T-DNA molecules in genetic engineering.

8.3 CO-RESIDENT TI- AND AT-PLASMID VECTORS

Agrobacterium tumefaciens's genetic composition consists of a circular and linear chromosomal DNA, the tumour inducing (Ti) and the *A. tumefaciens* (At) plasmid DNA. As indicated above, the process of introducing *goi* into *Agrobacterium* and plants starts with a complex microbial genetic modification that inserts these preferred sequences of genes into the T-DNA region of the Ti-plasmid vector (Gelvin 2003). But, to enhance the transient and stable transformation of plants with Ti-plasmid in collaboration with the helper plasmid (At), efforts have been made by researchers to remove oncogenic characteristics used by this bacterium for tumour development to make a binary vector system. In this case, the T-DNA region and the *vir* genes required for DNA processing and transfer were split into two replicons to form a binary system that efficaciously effects T-DNA integration in new hosts. Achieving this binary system has permitted and opened up the field of genetic engineering to numerous laboratories across the globe (Lee and Gelvin 2008).

Furthermore, a strategy to upregulate virulence genes expression levels has also increased the T-DNA transfer efficiency. Several reports have also indicated that the use of ternary systems for the application of *vir* gene inducers, changes in the origin of replication (Ori) of the binary vector and the use of super-binary vector systems as some of the subsequent improvements also allowed for the increase of the transformation frequencies (Ye et al. 2011; Hu et al. 2013; Nonaka et al. 2019). Tumour-inducing genes or oncogenes are disarmed and made non-functional by interrupting their sequences, especially to influence genetic transformation in plants. All these molecular modifications together with the development of the binary vector system have by far improved genetic engineering, including overcoming the problems caused by the larger sizes of the Ti-plasmids.

However, the strategy was based on the fact that *vir* genes functions are provided by a one disarmed Ti-plasmid in which the T-DNA and its left/right imperfect 25 base pair direct repeat border sequences have been deleted, leaving the oncogenes and other plasmid parts intact. The second plasmid, which is smaller (less than 200 base pair), contains an Ori, modified T-DNA between the abovementioned flanking border regions, the *goi* and antibiotic resistant genes used as selectable markers. Earlier reports showed that when the T-DNA and the *vir* region were present on the same plasmid, the term 'unitary' vectors was used, in contrast to 'binary' vectors where these two occurred in separate Ti- and At-plasmids (Spielmann and Simpson 1986).

8.4 GENE TRANSFER AND EXPRESSION

One of the most important benefits of using *Agrobacterium tumefaciens* compared to any other microbe has been its ability to deliver DNA to plant cells and permanently change the plant genome complement. This ability has been improved, enabling researchers to genetically transform plants for both agricultural purposes and basic research. Various research groups have then used disarmed *A. tumefaciens* vectors to transfer and express different segments of *goi* to generate various herbicide- and stress-resistant transgenic varieties that are currently in use. Several transformants containing genes such as *Bacillus thuringiensis endotoxin* gene, *beta glucuronidase* gene and *neomycin phosphotransferase* gene can be transferred and expressed, accompanied by a vastly different genetic transformation efficiency, especially in cultivars of the same species or genus. Although target tissues and species can be transformed, there is still a relatively poor understanding of the genetic transfer factors underlying these differences in T-DNA expressions.

In *Cleome gynandra*, which is a C_4 species of the Brassicales, Newell et al. (2010) reported a 14% *goi* expression efficiency using leaf tissues with significant detectable amounts of *β-glucuronidase* activity. By contrast with other Brassicales, 6.4 and 10% transformation frequencies were obtained using several lines of broccoli and cole cabbage, respectively. Variations in DNA expressions were obtained among all the broccoli lines of the same species. Likewise, Hisano and Sato (2016) compared different barley lines (*Hordeum vulgare*) to study the variations in gene expression that occurred among cultivars of the same species. This study identified significant segregation distortions of alleles that were attributed to the high transformation efficiencies. These conserved genomic sequences were identified in subsequent generation, and were termed Transformation Amenability (TFA) factors that potentially contained genetic regions that promoted *Agrobacterium*-mediated T-DNA expressions. Unfortunately, this evidence means that gene transfer and expression through *Agrobacterium*-mediated genetic transformation is still limited by both technical and molecular challenges in many recalcitrant crops.

These situations pose great challenges for researchers to focus their resources on developing a method of choice for plant transformation and dealing with the existing/emerging procedural limitations instead of assessing some of the observable functions and implications of the expressed genes on the mutant organisms as well as the environment.

8.5 INFECTION AND CO-CULTIVATION FOR EFFICIENT TRANSGENESIS

Recent and current reports still show that more research is still being focussed and directed towards investigating the use of *in vitro* culture conditions for genetic transformation in soybean, particularly explant infection and co-cultivation conditions (Paz et al. 2004; Mangena and Sehaole 2020). As already discussed in this book, *in vitro* transformation systems are prone to the problem of tissue senescence, chimerism and reductions on totipotent potential of targeted tissues. However, advancements in the establishment of an efficient *in vitro* regeneration system for genetic improvement is dependent on the type of explants used in a culture. However, cotyledonary node systems of transformation with or without pre-existing meristems have been widely reported in the production of genetically improved crop varieties (Li et al. 2017; Mangena 2020).

Most Gram-negative bacteria, including *Agrobacterium tumefaciens*, produce N-acyl homoserine lactones called autoinducers which act as diffusible intercellular pheromones. Intercellular pheromones are chemicals produced by microorganisms to change the behaviour of the host organism that they infect. When secreted or excreted by wild-type *A. tumefaciens*, these chemicals trigger the bacterium's pathogenicity against plants by regulating the genes responsible for plant defence response and immunity (Fuqua and Winans 1996). Response genes are activated via cognitive regulatory proteins with a sequence homologous to the LuxR, a transcriptional activator of bio-luminescence (*Lux*) gene found in *Vibrio fischeri*. The luxR-LuxI-type regulators are said to activate the expression of target genes preferentially at high cellular densities of bacteria and as a result they provide the bacteria with a mechanism for sensing their own population density (Zhu et al. 1998).

A. tumefaciens uses the LuxR-LuxI-type regulatory system to regulate the conjugal transfer of its Ti-plasmid using opines, especially octopines. *Agrobacterium* controls the release of octopines through the OccR activator to regulate the traR which is also homologous to LuxR (Fuqua and Winans 1996). In this case, traR is then used to regulate the expression of Ti-plasmid conjugal transfer genes in the presence of bacterial N-acyl homoserine lactones inducers. The mechanism of Ti-plasmid conjugal transfer is coupled with some of the receptor mechanisms from the host cells, including proteins like Attl, ChvA, ChvB and PscA. Plant host cells are triggered to induce receptor proteins such as the cellulase synthases CSLA9, ricadhesin binding protein, vitronectin protein and other proteins that interact with the VirB2 protein of *Agrobacterium* to effectively facilitate plant tissue transgenesis (Citovsky et al. 2007).

The above molecular processes contribute to the unique abilities of this bacterium to hijack critical cellular processes of the hosts during plant transformation. Therefore, for effective co-cultivation and transgenesis, *A. tumefaciens* requires activation of the activity of virulence genes, including T-complex and host cellular proteins. Both sets of regulatory gene products are highly required for host cell recognition, attachment and T-strand processing. Those molecular processes play a key role in nuclear import, integration and expression of the T-DNA taking place during the co-cultivation stage in the cells of the host plants.

8.6 MOLECULAR ROLE OF ORGANIC SUPPLEMENTS

Agrobacterium-mediated transformation involves multiple steps before fertile transgenic plantlets can be recovered. Sensing of host tissue wound sites until the T-DNA is expressed requires the deployment of a large number of proteins to enable the usage of host molecular machinery to initiate and execute this exogenous DNA transfer, integration and expression. While *Agrobacterium* attachment to the host cell is critically an important prerequisite step, under plant tissue culture, additives like acetosyringone help sense and attract the bacteria to the wound serving as an infection site (Xi et al. 2018). Moreover, this natural phenolic compound also induces virulence genes that facilitate the transfer of the T-DNA region into host plant cells. Acetosyringone, L-cysteine and dithiothreitol are some of the most important factors that facilitate the efficient delivery and integration of transgenes into the plant nuclear genome and by also improving plant regeneration *in vitro*.

When *Agrobacterium* infects plants to transfer a well-defined fragment of its genome (T-DNA), it may induce oxidative stress (Mangena 2021). Therefore, the addition of factors such as dithiothreitol helps to minimise the effects of tissue senescence during transgene integration after perception and attachment to host cells. *Agrobacterium tumefaciens* strains bear the genes that encode a LuxI-type (TraI) protein for the synthesis of N-(3-oxooctanoyl)-homoserine lactones, and LuxR-type (TraR) that binds to the 3-oxo-c_8 homoserine lactones. The expression of genes responsible for the release of these autoinducers is regulated by octopines related to plasmids like pTi15955, pTiA6 and pTiR10, and agrocinopines A and B associated with the induction of genes in the pTiC58 plasmid. It is now well known that the processing, transfer and integration of T-DNA are controlled by the activity of the virulence genes which are tightly regulated by the release of these chemicals.

Chemical signals such as octopines controlled by *Agrobacterium* and phenolic compounds like acetosyringone released from wounded plant tissues indeed serve as stimulus for *Agrobacterium* attachment and processing, as well as the translocation of the T-DNA molecule. These molecular processes include the exportation of the T-complex, along with other virulence proteins such as VirD5, VirE2, VirE3 and VirF that will mediate nuclear import of the T-strand after being localised with host cell cytoplasm (Pitzschke and Hirt 2010; Gelvin 2017).

8.7 DNA REPAIR FOR EFFICIENT INTEGRATION

The T-DNA located on the Ti/Ri-plasmid is transferred from *Agrobacterium* into the host cells as a single-stranded DNA. This DNA is defined by left and right border sequences that help to process the T-DNA from the Ti-plasmid (Gelvin 2003). The border sequences serve as an attachment site for the VirD2 protein that functions to direct the T-DNA to the plant genome, especially one found on the C-terminal part of the protein (Rossi et al. 1993). In addition to these, sequences serve as target sites for VirD1/VirD2 endonucleases for cleaving the T-region away from the Ti-plasmid when successful transfer into host cells has taken place (Yanofsky et al. 1986). It is these left and right border sequences that are annealed to the host genome during transgene integration. The host genome is also nicked where T-DNA is inserted and then ligated. According to Citovsky et al. (2007), this process is followed by a DNA repair process involving genome reconstruction known as gap-repair used for synthesis of a complementary strand converting the T-DNA into a double-stranded DNA molecule.

Therefore, the double-stranded breaks in the host DNA and converted double-stranded T-DNA intermediates found in the host genome play a vital role in the transgene integration and expression process. In yeast cells (*Saccharomyces cerevisiae*) *Agrobacterium* infection revealed that specific DNA repair genes that code for Ku70, MreII, Lig4, Rad50 and Sir4 proteins were responsible for T-DNA integration via the non-homologous DNA recombination pathway (van Attikum and Hooykaas 2003). This is, thus, probably a very similar mechanism used by plants during *Agrobacterium*-mediated genetic transformation. However, it is now known that major variations exist in the nature of somatic/germ-line cell transformation reflected by fundamental differences between host factors required for T-DNA integration, especially in somatic cells and in germ-line cells.

8.8 IMMUNITY TO PLANT TRANSFORMATION

Ample evidence is available indicating that T-DNA integration may require the role of both DNA repair proteins and DNA-packaging proteins like histones H2A to achieve higher expression levels of transgenesis inside the host cells. But plants may still perceive *Agrobacterium tumefaciens* and its transferred DNA molecules as foreign invaders that must be defended against and its T-complex removed from the host cell system. In this instance, plant cells may use their defence system to battle against the *Agrobacterium* infection process and expression of the T-DNA (Citovsky et al. 2007). Generally, plants defend themselves against pathogens (necrotrophs and biotrophs) by preformed structures and chemicals or through inducing an immune response after the infection. If the preformed structural mechanical barriers are successfully bridged by invading pathogens, plant receptors will initiate signalling pathways driving the expression of defence response genes.

In a review, Agier et al. (2018) demonstrated that plant defence systems rely upon pathogen direct detection of pathogenic molecules using pattern recognition receptors (PRRs) found within the cell membranes. This includes various other responses such as hypersensitive response (HR), ROS production, cell wall modifications, stomatal closure and the production of chemicals (proteins, chitinases, protein inhibitors, defensins and phytoalexins) that serve as antipathogenic chemicals. In turn, Gram-negative bacteria have also evolved a mechanism to produce indispensable ubiquitous polysaccharide components such as lipopolysaccharides (LPSs) on their cell wall surfaces to battle against plant-derived antimicrobial chemicals. LPSs have diverse roles during the pathogenesis of plants, but they exclusively and mainly promote bacterial growth to infect plants by altering PRR gene expression and prevent HR induction (Dow et al. 2000).

Similarly, it was observed that perhaps *A. tumefaciens* possesses this kind of an immune response inhibitors in plants. The *Agrobacterium*-mediated transformation of *Ageratum conyzoides* cells led to a reduced immune response of the plant. This study also revealed that the transformation efficiency of recovered plant cells was inversely proportional to the plant's immune response (Ditt et al. 2005). More findings that were in agreement with the above report were also highlighted by Citovsky et al. (2007) in *Arabidopsis* and tobacco. The plant response system was triggered by *A. tumefaciens* inoculation a few hours after infection to try to suppress infectious bacterial spread in the cells of the plant. Evidence of these kinds of responses has previously led to more investigations on host reactions towards *A. tumefaciens* infection and co-cultivation under laboratory conditions.

The following: (1) production of ROS through the perception of pathogen-associated molecular patterns (PAMPs), (2) formation of callus cells on infected explants to reinforce cell wall structures and (3) accumulation of salicylic acid through systemic acquired resistance (SAR) together with the (4) inactivation of pathogenesis-related genes, were reported as some of the pathways by which plants confer immunity against *A. tumefaciens* during *in vitro* genetic transformation (Durrant and Dong 2004). Other studies, however, reported counter measures evolved by *Agrobacterium* in evading plant immunity. Among some of these measures, the inhibition of SAR through the reduction of *PR1* and *PR5* gene expression levels was identified. PRs refers to pathogen-related gene product proteins classified into 17 families involved in the accumulation of defence-related metabolites and proteins (van Loon et al. 1994).

8.9 SUMMARY

Molecular analysis of *Agrobacterium tumefaciens*-mediated genetic transformation has so far revealed and supported the development and implementation of numerous *in vitro* and *in vivo* protocols used in the improvement of many monocot and dicot species (Mangena and Sehaole 2020). This analysis also revealed that the genetic engineering process is internally divided into several critical stages. The stages first involve a complete genetic engineering of the transforming bacterium itself before T-DNA can be translocated into host plant cells and tissues. Although *Agrobacterium* has a complete transgene transfer and integration machinery, the mechanism will not function effectively without the involvement of the host plant's molecular processes, particularly for the development of highly efficient transformation strategies which are still lacking. Furthermore, pathogenesis by this soil-borne Gram-negative proteobacterium still faces several challenges in its host-bacterium associations, virulence factors and the problem of plant immunity.

To date, some studies have reported that *A. tumefaciens* possesses different mechanisms that may help the bacterium in effectively dealing with issues relating to plant immune and resistance responses. Such pathways include the bacterium's ability to change patterns of pathogen recognition receptor genes, use of LPSs and reducing the expression levels of PR genes in the hosts. Future advances in the study of molecular mechanisms involving pathogenesis of both germ/reproductive and somatic tissues in the acquisition of competent cell lines for efficient *in vitro* plant transformation should focus on such above mentioned challenges. Such endeavours may give further useful insights that may contribute to our understanding of genetic engineering in soybean using *A. tumefaciens*, especially at the molecular level. Molecular processes serve as the smallest and most complex level in studying the structure, function and role of each genomic component contributing to the immunity or susceptibility of host plant cells to *Agrobacterium* transformation.

8.10 ABBREVIATIONS

At-plasmid *Agrobacterium tumefaciens* plasmid
BNA Brazil nut 2S albumin
DNA Deoxyribonucleic acid
Goi Gene of interest

HVA1	*Hordeum vulgare* abundant protein 1
Met	Methionine
Ri-plasmid	Hairy root-inducing plasmid
T-DNA	Transferred-DNA
TFA	Transformation amenability
Ti-plasmid	Tumour-inducing plasmid
T4SS	Type 4 secretion system
Vir	Virulence

REFERENCES

Agier J, Pastwiska J and Brzazinska-Braszczyk E. (2018). An overview of mast cell pattern recognition receptors. *Inflammation Research* 67, 737–746.

Barker RF, Idler KB, Thompson DV and Kemp JD. (1983). Nucleotide sequence of the T-DNA region from the *Agrobacterium tumefaciens* octopine Ti-plasmid pTi15955. *Plant Molecular Biology* 2, 335–350.

Citovsky V, Kozbvsky SV, Lacroix B, Zaltsman A, Dafny-Yelin M, Vyas S, Tovkach A and Tzfrira T. (2007). Biological systems of the host cell involved in *Agrobacterium* infection. *Cell Microbiology* 9(1), 9–10.

Ditt RF, Nester E and Comai L. (2005). The plant cell defense and *Agrobacterium tumefaciens. FEMS Microbiology Letters* 247, 207–213.

Dow M, Newman M-A and von Roepenack E. (2000). The induction and modulation of plant defense responses by bacterial lipopolysaccharides. *Annual Review of Phytopathology* 38, 241–261.

Durrant WE and Dong X. (2004). Systemic acquired resistance. *Annual Reviews of Phytopathology* 42, 185–209.

Fuqua C and Winans SC. (1996). Conserved cis-acting promoter elements are required for density-dependent transcription of *Agrobacterium tumefaciens* conjugal transfer genes. *Journal of Bacteriology* 178(2), 435–440.

Gelvin SB. (2003). *Agrobacterium*-mediated plant transformation: The biology behind the "gene jockeying" tool. *Microbiology and Molecular Biology Reviews* 67(1), 16–37.

Gelvin SB. (2017). Integration of *Agrobacterium* T-DNA into the plant genome. *Annual Reviews of Genetics* 51, 195–217.

Hinchee MAW, Connor-Ward DV, Newell CA, McDonnell RE, Sato SJ, Gasser CS, Fischhoff DA, Re DB, Fraley RT and Horch RB. (1988). Production of transgenic soybean plants using *Agrobacterium*-mediated DNA transfer. *Biotechnology* 6, 915–921.

Hisano H and Sato K. (2016). Genomic regions responsible for amenability to *Agrobacterium*-mediated transformation in barley. *Scientific Reports* 6(37505), 1–11.

Hu X, Zhao J, DeGrado WF and Binns AN. (2013). *Agrobacterium tumefaciens* recognises its host environment using ChvE to bind diverse plant sugars as virulence signal. *Proceedings of the National Academy of Science of the United State of America* 110(2), 678–683.

Hwang H-H, Yu M and Lai E-M. (2017). *Agrobacterium*-mediated plant transformation: Biology and applications. *The Arabidopsis Book* 15(e0186), 1–32.

Jain A and Srivastava P. (2013). Broad host range plasmids. *FEMS Microbiology Letters* 348(2), 87–96.

Kahrstrom CT. (2014). Solving the T4SS structural mystery. *Nature Review Microbiology* 12, 312–313.

Ko TS, Nelson RL and Korban S. (2004). Screening multiple soybean cultivars (MG00 to MGVIII) for somatic embryogenesis following *Agrobacterium*-mediated transformation of immature cotyledons. *Crop Science* 44, 1825–1831.

Kwapata K, Nguyen T and Sticklen M. (2012). Genetic transformation of common bean (*Phaseolus vulgaris* L.) with the *Gus* color marker, the *bar* herbicide resistance and the barkey (*Hordeum vulgare*) HVA1 drought tolerance genes. *International Journal of Agronomy* Article ID198960, 1–8.

Lee L-Y and Gelvin SB. (2008). T-DNA binary vectors and systems. *Plant Physiology* 146(2), 325–332.

Li S, Cong Y, Liu Y, Wang T, Shuai Q, Chen N, Gai JJ and Li Y. (2017). Optimization of *Agrobacterium*-mediated transformation in soybean. *Frontiers in Plant Science* 8(246), 1–15.

Mangena P. (2020). Genetic transformation to confer drought stress tolerance in soybean (*Glycine max* L.). In Guleria P, Kumar V and Lichtfouse E (eds), *Sustainable Agriculture Reviews. Sustainable Agriculture Reviews*, Vol 45. Springer, Cham. pp. 193–224.

Mangena P. (2021). Effect of *Agrobacterium* co-cultivation stage on explant response for subsequent genetic transformation in soybean (Glycine max (L.) Merr.). *Plant Science Today* 8(4), 905–911.

Mangena P and Sehaole EKM. (2020). Transgenic grain legumes. In Mangena P (ed), *Advances in Legume Research: Physiological Responses and Genetic Improvement for Stress Resistance*, Vol 1. Bentham Science Publishers, Sangapore. pp. 148–172.

McCullen CA and Binns AN. (2006). Agrobacterium tumefaciens and plant cell interactions and activities required for interkingdom macromolecular transfer. *Annual Review of Cell and Developmental Biology* 22, 101–127.

Mehrotra M, Sanyal I and Amla DV. (2011). High-efficiency *Agrobacterium*-mediated transformation of chickpea (*Cicer arietinum* L.) and regeneration of insect-resistant transgenic plants. *Plant Cell Report* 30(9), 1603–1616.

Muntz K, Saalbach VCG, Waddell ISD, Schieder TPO and Wustenhagen TC. (1998). Genetic engineering for high methionine grain legume. *Hahrung* 42(2–4), 125–127.

Newell CA, Brown NJ, Liu Z, Pflug A, Gowik U, Westhoff P and Hibberd JM. (2010). *Agrobacterium tumefaciens*-mediated transformation of *Cleome gynandra* L., a C_4 dicotyledon that is closely related to *Arabidopsis thaliana*. *Journal of Experimental Botany* 61(5), 1311–1319.

Nonaka S, Soweya T, Kadota Y, Nakamura K and Ezura H. (2019). Super-*Agrobacterium* ver. 4: Improving the transformation frequencies and genetic engineering possibilities for crop plants. *Frontiers in Plant Science* 10(1204), 1–12.

Patel M, Dewey RE and Qu R. (2013). Enhancing *Agrobacterium tumefaciens*-mediated transformation efficiency of perennial ryegrass and rice using heat and high maltose treatments during bacterial infection. *Plant Cell, Tissue and Organ Culture* 114, 19–29.

Paz MM, Shou H, Guo Z, Zhang Z, Benerjee AK and Wang K. (2004). Assessment of conditions affecting *Agrobacterium*-mediated soybean transformation using the cotyledonary node explant. *Euphytica* 136, 167–179.

Pitzschke A and Hirt H. (2010). New insights into an old story: *Agrobacterium*-induced tumore formation in plants by plant transformation. *EMBO Journal* 29, 1021–1032.

Rossi L, Hohn B and Tinland B. (1993). The virD2 protein of *Agrobacterium tumefaciens* carries nuclear localization signals important for transfection of T-DNA to plant. *Molecular and General Genetics* 239(3), 345–353.

Song QQ, Han X, Wiersma AT, Zong X, Awale HE and Kelly JD. (2020). Induction of competent cells for *Agrobacterium tumefaciens*-mediated stable transformation of common bean (Phaseoulus vulgaris L.). *PLOS One* 15(3), e02299909, 1–16.

Sood P, Bhattacharya A and Sood A. (2011). Problems and possibilities of monocot transformation. *Biologia Plantarum* 55(1), 1–15.

Spielman A and Simpson RB. (1986). T-DNA structure in transgenic tobacco plants with multiple independent integration sites. *Molecular and General Genetics* 205, 34–41.

van Attikum H and Hooykaas PJJ. (2003). Genetic requirements for the targeted integration of *Agrobacterium* T-DNA in Saccharomyces cerevisiae. *Nucleic Acids Research* 31, 826–832.

van Loon LC, Pierpoint WS, Voller T and Conejero V. (1994). Recommendations for naming plant pathogenesis- related protein. *Plant Molecular Biology Reporter* 12, 245–264.

Xi J, Patel M, Dong S, Que Q and Qu R. (2018). Acetosyringone treatment duration affects large T-DNA molecule transfer to rice callus. *BMC Biotechnology* 18(48), 1–8.

Yanofsky MF, Porter SG, Young C, Albright LM, Gordon NP and Nester EW. (1986). The virD operon of *Agrobacterium tumefaciens* encodes a site-specific endonuclease. *Cell* 47(3), 471–477.

Ye X, Williams EJ, Shen J, Johnson S, Lowe B, Radke S, Strickland S, Esser JA, Petersen WM and Gilberton LA. (2011). Enhanced production of single copy backbone-free transgenic plants in multiple crop species using binary vectors with a pRi replication origin in *Agrobacterium tumefaciens*. *Transgenic Research* 20 (4), 773–786.

Zhang Z, Xing A, Staswick P and Clemete T. (1999). The use of glufosinate as selective agent in *Agrobacterium*-mediated transformation of soybean. *Plant Cell, Tissue and Organ Culture* 56, 37–46.

Zhu J, Beaber JW, More MI, Fuqua C, Eberhard A and Winans SC. (1998). Analogs of the autoinducer 3-oxooctanoyl-homoserine lactone strongly inhibit activity of the TraR protein of *Agrobacterium tumefaciens*. *Journal of Bacteriology* 180(20), 5398–5405.

9 Genetic Transformation in Other Leguminous Crops

9.1 INTRODUCTION

The globe has entered into the expansion phase of commercial agriculture, in which genetically modified crops are highly required to curb productivity challenges, food shortages, poverty and hunger. These are much needed efforts, particularly in developing countries. Legume species such as common bean (*Phaseolus vulgaris*), chickpea (*Cicer arietinum*), pigeon pea (*Cajanus cajan*), mung bean (*Vigna radiata*), cowpea (*Vigna unguiculata*), faba bean (*Vicia faba*), peas (*Pisum sativum*) and lentil (*Lens culinaris*) in addition to soybean serve among the edible food crops, directly or indirectly in the form of ripe or unripe seeds. Most of these cultivated legumes are used for industrial processing to develop and produce products used in the manufacturing and pharmaceutical industries. Many of these species belong to the *Vicieae* and *Phaseolus* tribe as indicated in Table 9.1. The species exhibit phylogenetic characters showing a strong relationship based on their evolutionary development and diversification that are found within the Fabaceae family (Mangena 2020).

Phylogenetic characters highlighted in Table 9.1 indicate a clear inherent genetic control demonstrating the fact that many species within these tribes are restricted to epigeal or hypogeal kinds of germination, together with either determinate or indeterminate growth forms. Determinate form refers to a type of growth where the main stem growth ends in an inflorescence or other reproductive structures. Meanwhile, an indeterminate growth form shows the continuation of stem elongation indefinitely. These biological characters serve as key diagnostic features that are a representation of residual traits that have evolved from the plants' ancestral associates (Lavin et al. 2001). Much of the biological phylogenetic characters are targeted and exploited for genetic improvement, especially for the eight legume species mentioned above, including soybean (*Glycine max* (L.) Merrill). Furthermore, many hybrid lines derived from these species have dramatic improvements in growth and yields. However, challenges still remain, both in the production of better performing hybrids and improvement of the technologies which have the potential to address many of these challenges.

This chapter, therefore, provides a brief overview of the use of *Agrobacterium tumefaciens*-mediated genetic transformation in leguminous crops of commercial importance. Many of these crop species possess grain qualities that are suitable for human and animal consumption, as well as for industrial processing. They have traditionally, to date, occupied a crucial position in legume-based diets to serve the

DOI: 10.1201/b22829-9

TABLE 9.1
Botanical names and common names of some of the grain legumes and their tribes, found under the family Fabaceae

Tribe	Species	Tribe phylogenetic/taxonomic characters
Vicieae/ Fabeae	*Lens culinaris* Medik. *Pisum sativum* L. *Vicia faba* L. *Vicia ervilia* (L.) Willd.	• Solitary or racemose, pubescent flowers, • Paripinnate leaves, with tendrils, • Well-developed stipules, • Usually low or climbing herbs.
Cicereae	*Cicer arietinum* L.	• Pink flowers, with anthocyanin pigmented stems, • White or beige-coloured seeds with ram's head shape.
Dalbergieae	*Arachis hypogeae* L.	• Plestomorphic flowers (free keel petals, staminal filaments partly fused), • Filaments without basal fenestrae, • Inflorescence of determinate growth.
Genisteae	*Lupinus luteus* L *Lupinus albus* L.	• Hilum positioned laterally with straight hilar groove, • Pods oblong or linear-oblong, with transversely ovate seeds, • Small ovaries producing larger pods.
Indigofereae	*Cyamopsis tetragonoloba* (L.) Taub	• Presence of re-carmine flowers, lacking anther hairs, • Perforated pollen, • Paniculate inflorescences.
Phaseoleae	*Canavalia gladiate* (Jacq.) *Canavalia ensiformis* (L.) *Cajanus cajan* (L.) Huth *Glycine max* (L.) Merrill *Mucuna pruriens* (L.) *Phaseolus acutifolius* *Phaseolus lanatus* L. *Phaseolus vulgaris* L. *Vigna radiata* (L.) Wilczek *Vigna aconitifolia* (Jacq.) *Vigna mungo* (L.) *Vigna umbellate* (Thumb.) *Vigna unguiculata* (L.) Walp	• Style tip is expanded and spoon, • Bearded style, • Standard petals with appendages, • Hilum covered with white spongy tissues.

nutritional, health and welfare needs of many populations as vital sources of proteins, minerals, carbohydrates and vitamins. It is for some of these critical reasons that intensification of legume agriculture should be accompanied by major enhancements in their growth, yield and stress resistance characteristics.

9.2 STATUS OF GENETIC TRANSFORMATION IN OTHER LEGUME CROPS

Adverse impacts of abiotic stress factors such as chilling, drought, heat stress, etc., and the spread of diseases warranted and demanded a rapid response towards the adoption of genetically engineered legume crops. As already stated, these are transgenic lines that allow for the expression of exogenous coding sequences such as transcription factors, reporter genes, resistance genes and others that confer tolerance to biotic and abiotic stress to improve the growth and yield of crops. The characterisation of these exogenous DNA shuttle mechanisms which facilitate transgene transfer, integration and expression in plant host cells has been an active area of research involving legumes for decades. Fortunately, a practical outcome of this complex mechanism is that Ri/Ti-plasmids carrying the T-region with the gene of interest have been successfully imported and robustly expressed in different legume species for the efficient production of transgenic plants. Some of the specific successes and challenges of those efficiently introduced exogenous DNA molecules into grain legume crops are briefly discussed below. Although other major breakthroughs may have been well established and achieved through different gene transfer methods, the current discussion will only focus on *Agrobacterium tumefaciens*-mediated genetic transformation of the below identified grain legume crops.

9.2.1 Chickpea – *Cicer arietinum*

Chickpea is one of the most important leguminous, cool-season, food legumes and is predominantly cultivated in the Asian Pacific region. Chickpea is also grown in other parts of the world including sub-Saharan Africa, Europe, North America and South America for its nutritional value. It contributes about 15% to the world's pulse harvest, which is about 58 million tons annually. Agricultural production of chickpea has been very low compared to other legumes due to various refractory biotic and abiotic stress factors (Sharma et al. 2006). Abiotic stresses like chilling and drought stress, as well as biotic constraints such as *Ascochyta* blight (AB), *Botrytis* grey mould (BGM), *Fusarium* wilt and pod borer deter successful plant growth and cause higher yield losses.

Agrobacterium tumefaciens gene transfer is one of the methods tested to confer effective resistance against pests, diseases and abiotic stress. This protocol has resulted in notable interchanges of desirable traits that are important to pulse crops. An efficient *Agrobacterium* protocol for chickpea using axillary meristem explants was reported (Sharma et al. 2006). This protocol resulted in 70% frequency of transformation and successful recovery of valuable transgenic plants. A co-transformation frequency of *cry1Ac* and *nptII* of about 1.12% was achieved. This followed the expression of a 2.2 kb native truncated *Bacillus thuringiensis* (*Bt*) *cry1Ac* gene driven by the CaMV35S promoter and the nptII gene for kanamycin selection in cotyledonary nodes after the preconditioning of explants with *A. tumefaciens*, acetosyringone, dithiothreitol and L-cysteine during co-cultivation.

These were coupled with sonication and vacuum treatment for efficient genetic transformation (Sanyal et al. 2005). In another study, the infection of apical meristem explants with *Agrobacterium* strain EHA105 carrying the pCAMBIA2301 vector with *uidA* and *nptII* genes was conducted. Again, an overall chickpea transformation frequency of 1.2% was achieved (Srivastava et al. 2017). A high transformation efficiency between 23 and 61% was reported on chickpea roots with the expression of green florescent protein and *AtTT2* gene involved in proanthocyanidins biosynthesis

(Aggarwal et al. 2018). As the development of an efficient and reproducible transformation protocol is imperative to facilitate functional genomics in this crop, challenges such as genotype specificity and the production of phenolics upon excision and infection of explants with *A. tumefaciens* still persist.

High phenolic content expressed by tissues has proved detrimental to the availability of *A. tumefaciens* as indicated in previous chapters. It was also evident that the extent of endogenously induced hypersensitive response of plant tissue against *Agrobacterium* infestation was also genotype dependent. These factors inhibiting the production of stable transgenic plants in chickpea are similar to constraints influencing the overall role of plant transformation in recalcitrant legume genotypes.

9.2.2 Pigeon Pea – *Cajanus cajan*

Conventional and modern breeding technologies have resulted in limited successes in the genetic improvement of pigeon pea. This observation is similar to the limited genetic interchanges of desirable characters in cowpea that will be discussed later in this chapter. Unlike soybean, pigeon pea transformation rates were mostly found to be extremely low, and the process of transformation for this crop was also found to be very time consuming. In a recent study by Karmakar et al. (2019) an overall transformation efficiency of approximately 83% was reported using a 0.25 optical density of *Agrobacterium tumefaciens. Agrobacterium* cells with 15 minutes' and 72 hours' duration for bacterial inoculation and co-cultivation, respectively, were reported. Furthermore, this study used seeds of pigeon pea cultivar ICPL85063 that were imbibed in sterile distilled water for 16–18 hours before being used to develop cotyledon explants with the embryogenic axis still attached to them.

This study served as one of the most highly efficient and rapid protocols for the genetic transformation of pigeon pea, as well as legumes in general. Earlier reports such as those of Verma and Chand (2005) and Surekha et al. (2005), in contrast, achieved very low transformation frequencies also using explants derived from germinated seeds. Transformation efficiencies of 0.33 and 0.20% were recorded for *uidA* and *cry1A(b)* genes by optimised gene transfer procedure evaluating leaf disc, cotyledons and mature decapitated embryogenic axes of Bahar and UPASRO cultivars. Furthermore, most studies demonstrated that the use of embryogenic axes explants is so far the best explant type for the expression of exogenous DNA in this crop. Pigeon pea needs genetic improvements since it is used for food, fodder, firewood and building materials in a few Asian and African countries that use it. Like chickpea and many other crop plants, its growth and productivity are constrained by the various biotic and abiotic stress factors.

9.2.3 Lentil – *Lens culinaris*

Lentils are protein-rich grain legumes grown in semi-arid tropics. Lentil seeds are considered to contain high protein levels of about 26% on average, comparable to faba bean but higher than chickpea and even monocots like wheat (Akcay et al. 2009).

Like pigeon pea, genetic transformation studies in lentil are very limited, despite the high amounts of proteins they contain, their ability to boost health and other essential nutrients. Lentils, like any other legume plant, fix atmospheric nitrogen for soil enrichment. None of the several studies published before 2009 reported a stable formation and recovery of lentil transgenic plants using *Agrobacterium tumefaciens*-mediated genetic transformation or a microprojectile bombardment system. A few transformed plants that were phenotypically normal were earlier reported. These plants survived metsulfuron-methyl herbicide screening following mutant acetolactate synthase gene (ALS) bombardment.

Lentil cotyledonary nodes derived from seedlings germinated on MS medium containing 0.7% BAP were regenerated *in vitro* as explants for ALS bombardment (Gulati et al. 2002). Several other studies also reported regeneration frequencies ranging between 0 and 1% following *Agrobacterium* and particle bombardment mediated genetic transformation. These studies used shoot apex, cotyledonary nodes, epicotyls, nodal segments and roots as explants. A stable, simple and reproducible *Agrobacterium*-based gene transfer protocol for lentil was later developed for two microsperma varieties Bari Masur-4 (BM-4) and Bari Masur-5 (BM-5) using a strain LBA4404 containing a binary plasmid vector pBI121 with *GUS* and *nptII* marker genes. Different types of explants, namely decapitated embryo, cotyledon attached to these embryos and cotyledonary nodes, were used in this study to yield at least 1.009% overall frequency of genetic transformation.

The stable integration of transgenes was confirmed using *GUS* positive expression detection more in decapitated embryos, followed by embryos attached to cotyledons and then on cotyledonary node explants. Further confirmation of the results was done using PCR analysis following the selection of transgenic plants using elevating concentrations of kanamycin (Das et al. 2012). Most recently, a chitinase gene was successfully transferred and expressed in BM-4, BM-5 and BM-6 lentil varieties using *Agrobacterium* strain EHA105 containing the *bar* gene whereby the selection of transgenic plants was done using phosphinothricin (glufosinate) at 2.0 mg/l. A total of 0.36% was achieved for BM-4 and BM-6; meanwhile BM-5 recorded 0.34% transformation frequency (Das et al. 2019). Therefore, the most recent and current published studies involving lentil transformation clearly show that this crop legume species remains highly recalcitrant.

To date, none of the *Agrobacterium*-mediated genetic transformation studies have yielded an efficiency higher than 1% in both *in vivo*- and *in vitro*-based transformation culture experiments. Other techniques tested for this purpose such as electroporation and microprojectile bombardment were largely found to be very difficult to reproduce. However, lentil transformation lags far behind *Agrobacterium*-mediated gene transfer of soybean and other legume crops. Additionally, cotyledonary node explants developed from germinated seeds used in cowpea, faba bean, soybean and pigeon pea transformation are not the most common and predominant type of explants used for lentil transformation. In this case, the use of embryo axes has achieved higher frequencies of transformation than the single or double cotyledonary node explants utilised for genetic transformation.

9.2.4 Common Bean – *Phaseolus vulgaris*

Despite being among the earliest domesticated legume crops, first in Central Mexico which was followed by South America and then later across the globe about 8000 years ago, both traditional and modern breeding of common bean still pose several challenges (Castro-Guerrero et al. 2016). Currently, this crop is distributed and used as a staple food across the rest of the world, including Africa, Europe and Asia. Common bean together with soybean serve as the most important sources of proteins for human consumption and animal feed manufacturing. Both legumes play a critical role in the development of countries' economies, and have a very fascinating domestication history. Common bean is widely cultivated in developing countries due to its affordability, wider adaptation and tolerance to mild and sometimes transient severe drought stress (Barampama and Simard 1993). This crop is widely appreciated due to its low nutrient input during cultivation. However, common bean growth and yield can be severely reduced and negatively impacted by limited phosphate, iron and zinc found in the growing medium.

But the advances made in legume improvement also offer the opportunity to apply biotechnological techniques and approaches that allow for an exclusive genetic improvement of the common bean proactive response to biotic and abiotic stress constraints. Even though there is mixed abundance of common bean's gene pool across continents depending on the region, this crop still needs to be genetically improved in order to increase further its geographical distribution, especially for desirable agronomic traits to be selected for breeding purposes. Interestingly, common bean possesses a protein, phaseolina, which has structural properties similar to those of the globulins occurring in soybeans and other leguminous crops (Sabbah et al. 2019). Phaseolin is a major globulin found within different varieties of *Phaseolus vulgaris* (L.), and it is a trimeric, vicilin-like 7S polypeptide known to exhibit polymorphism of polypeptides. This protein also acts as an acyl donor and an acceptor when regulated by the activity of transglutaminase enzymes.

These cross-linking transferases catalyse the formation of covalent bonds between isopeptides in protein molecules that are normally derived from different sources (Sathe 2016). In total 7S phaseolin constitutes approximately 50% of the total seed proteins in common beans. This high content of essential proteins is important for nutrition, boosting attempts to genetically improve this crop for the sole purpose of enhancing its grain quality and bioavailability of such proteins. As a result, the selection and breeding of highly digestible phaseolin proteins is still being conducted (Montoya et al. 2010). Among the ways to improve nutritional quality, biotic and abiotic stress resistance in common bean, the testing of *Agrobacterium*-mediated genetic transformation in *P. vulgaris* is also ongoing. Many reports are available, particularly on the improvement against diseases and insect pests. The transformation of common bean (cv. Mwitemania and Rosa coco) was monitored using *GUS* expression on regenerated shoots and roots carrying intron plasmids.

A. tumefaciens strains LBA4404 (pB1121), EHA105 (pCAMBIA1201) and EHA105 (pCAMBIA1301) were evaluated for common bean transformation using 2-day-old half seed embryos as explants (Amugune et al. 2011). Recently, a 0.5% transfection efficiency was achieved using a cultivar Brunca. *Agrobacterium*-mediated gene transfer and expression were conducted using *gusA* and *nptII* genes, together with the trehalose-6-phosphate-synthase (*TPS1*) gene from *S. cerevisiae* (Solis-Ramos et al. 2019). Finally, common bean forms part of a group of recalcitrant legumes that lack efficient regeneration and transformation through *in vitro* culture protocols. Thus, to date, both indirect and direct gene transfer methods still show low transformation efficiencies ranging between 0.03 and 0.9%, including the use of *in vivo* approaches that bypass many *in vitro* regeneration barriers.

9.2.5 Peas – *Pisum sativum*

The first most stable transformation of peas was reported by Puonti-Kaerlas et al. (1990) more than three decades ago. In this study, Southern analysis of nuclear DNA from transformed pea materials revealed that all tested transgenic calli and plants had stable integration of the transgene. Pea cells were transformed with the GV3101 strain whereby both the callus cell DNA and DNA of plants regenerated from these calli cells gave very similar hybridization patterns. This study serves as one of the most efficient transformations using the hygromycin phosphotransferase (*hpt*) or *nptII* gene as a selectable marker. According to the report, the choice of the selectable marker, three step regeneration system and the use of explants derived from axenic shoot cultures were found to be of paramount importance for the high regeneration competence and subsequent genetic transformation. Like the rest of the recalcitrant legumes, transformation cultures and genotypes often yield varied results. Genotypes and cultivars as transformation factors continue to play a vital role in the susceptibility of legume crop species to the various *Agrobacterium tumefaciens* strains tested for genetic transformation.

Although several successes, including Puonti-Kaerlas' report, were recorded, the genetic transformation of peas remains a major challenge as in many dicotyledonous crops. Only a total of 5% shoots were proliferated from epicotyl segments and nodus explants established from etiolated seedling for genetic improvement purposes. Factors attributed to this poor response included explant source, *Agrobacterium* strain, co-cultivation condition/period and the pea genotype used (De Khatter and Jacobsen 1990). Although the problem of pea transformation was intensely discussed in review papers and in many other research platforms, these efforts still did not bring any positive outcomes. Among such efforts, the attempts at using direct gene transfer methods like particle bombardment were also unsuccessful (Svabova et al. 2005). The most promising protocols were based on organogenesis on callus or proliferation of organised meristematic tissues established through cotyledon, nodal or immature/mature embryo cultures. To date, most transformation protocols still do not represent a routine technology for rapid *Agrobacterium*-mediated genetic transformation in peas.

9.2.6 Faba Bean – *Vicia faba*

According to the recent literature, *Vicia faba* is one of the legume crops whose production of genetically transformed varieties has not been adequately reported. A relatively stable and feasible genetic engineering protocol of faba bean was reported using stem segment explants derived from aseptically germinated seedlings. *Agrobacterium tumefaciens* strain EHA105/101 containing a binary vector carrying a *uidA* gene, a mutant *lysC* gene and the coding sequence for a methionine-rich sunflower 2S albumin, together with a *nptII* as a selectable marker was used for transformation (Boettinger et al. 2001). Although the developed protocol was very cumbersome and lengthy, this may be considered the first successful study that identified an efficient *Agrobacterium*-mediated transformation of faba beans. A few years later, direct shoot organogenesis and transformed plantlet regeneration were achieved using embryogenic axes derived from apical meristematic cells as explants (Hanafy et al. 2005).

This study also co-cultivated the explants with *Agrobacterium* strain EHA105/pGIsfa carrying sulphur-rich 8S sunflower albumin coding sequence linked to the *bar* gene selectable markers. Compared to competing legume crops like peas and lentil, faba bean regeneration frequency as well as transformation rates were very low. This remain the case, despite faba bean ranking among the most important grain legumes and contributing more than 4.5 million tonnes on average to the global production statistics. According to the global pulse crops statistics projections for the year 2020–2021, this crop was ranked fifth under pulses and it is expected to expand by 4.6% of its global production (Janusauskaite and Razbadauskiene 2021; Bogale et al. 2021).

9.2.7 Cowpea – *Vigna unguiculata*

Cowpea is among the oldest and widely cultivated herbaceous legume crops in the sub-Saharan African countries. This crop is considered to be indigenous to Africa and serves as a major source of protein for both rural and urban dwellers in the continent. Cowpea leaves and pods are used for human consumption as vegetable and dried grain, respectively. The protein content ranges from 27 to 43% and 21 to 33% for leaves and dry seeds, respectively (Abudulai et al. 2016). However, cowpea yields can be adversely affected by a number of factors. Generally, the crop's growth and yields can be reduced due to the lack of improved breeding approaches, enhanced cultivars, poor agricultural inputs (fertilisers, agrochemicals, machinery and irrigation systems) and poor management practices (Sheahan and Barrett 2017). Agricultural practices such as intercropping of cowpea with maize, cassava and sorghum commonly practised in West and Southern Africa lead to very low grain yield of approximately 275 kgha^{-1} as a result of poor planting arrangement. Intercropping causes shading by companion crops and low plant population (Kyei-Boahen et al. 2017).

Among the many negative factors that influence cowpea productivity are soil fertility, pest and diseases coupled with the continued use of low-yielding disease-susceptible varieties. The continuous cropping of unimproved varieties like in any other crop definitely leads to a progressive decline in the growth and yield of cowpea. Therefore, addressing this problem may require massive changes in its breeding systems, especially to discontinue the planting of low-yielding varieties with poor plant establishment and growth. These would include the use of *Agrobacterium*-mediated genetic transformation as one of the precise tools for gene transfer and the development of high-yielding transgenic plants. Bett et al. (2019) reported the expression of transgene encoding an insecticidal protein from *Bacillus thuringiensis* (*Bt*) in cowpea for biotic stress resistance.

This study alternated two selection regimes using kanamycin and geneticin following a sonication procedure of *A. tumefaciens* AGL1 with a *Vip3Ba* gene construct to transform cotyledonary axes which still had radicle tips attached to them. About 1% transformation efficiency was recorded for transgenic plant resistance against pest insects responsible for the majority of cowpea yield losses in agriculture. Insect pests have been managed by spraying synthetic and natural agrochemicals that have proved to be very costly, health threatening and environmentally unsafe. The use of recombinant proteins encoded by gene sequences that are introduced into cowpeas through *Agrobacterium*-mediated genetic transformation may be the most feasible and effective way of dealing with problems such as pod borers, thrips and beetles. Furthermore, this recombinant DNA technology may also be used to complement the crossing of cowpea with other incompatible species such as *Vigna vexillate*. This species inherently contains resistance genes against *Maruca vitrata* causing bean pod borer that is not present in cowpea (Bett et al. 2019).

9.2.8 Mung Bean – *Vigna radiata*

Mung bean is an annual legume crop that is primarily cultivated for sprout production and cake flour. This crop is highly rich in carbohydrates (60%) constituting an amylose of about 56%, protein, fats, vitamins and minerals. Amylose makes mung bean characteristically resistant to cooking and gives it a smooth taste when commonly prepared as noodles. Mung bean noodles are one of the most common foods in Central China and Australia. Noodle preparation mainly includes grinding, fermentation, mixing, extrusion, boiling, cooling and drying (Tang et al. 2014). Mung bean also contains health-promoting effects in addition to the good nutritional value, and these are effects such as cholesterol-lowering, antiallergy, antibacterial and antitumour effects (Hou et al. 2019). With increasing nutritional and health benefits, consumption of mung bean has been rapidly growing along with some of its other functions like its use in medicine and cosmetics manufacturing.

On the other hand, the development of mung bean varieties that have a short growth cycle (recommended duration of about 70–90 days) may pave way for the expansion of its production and services in regions like South America and sub-Saharan Africa. One advantage is that this crop is relatively drought tolerant and requires low agricultural inputs. Furthermore, improved varieties must also be able to deal with the various biotic and other abiotic stressors that may still negatively impact on their growth and productivity. Insect pests (aphids, bruchids, whitefly and pod borer) and diseases (yellow mosaic, anthracnose, powdery mildew, halo blight, bacterial leaf spot and tan spot) are the major biotic stressors. Key abiotic stress constraints for mung bean include waterlogging, salinity stress and heat stress (Nair et al. 2019). These factors limit the growth and yield of mung bean, indicating that breeding is critical to develop varieties with improved resistance.

However, progress in this regard has been very slow since there is no precise and accurate identification of resistant gene sources for some of the traits, together with traits that are expressed by multiple genes as indicated by Nair et al. (2019). Unlike soybean, cowpea and faba bean, *Agrobacterium*-mediated genetic transformation of mung bean has been successfully achieved. Molecular analysis of putative transformed plants revealed the integration and expression of transgenes in the parent generation (T_0) and their seeds (T_1–T_3) following the transformation of hypocotyl and cotyledonary node (primary leaves) explants with *Agrobacterium* strain LBA4404 (pTOK233), EHA105 (pBin9GusInt) and C58C1 (pIG121Hm) also containg *gusA* and *nptII* maker genes (Jaiwal et al. 2001). Other studies that showed highly successful and efficient genetic transformation of this crop through *A. tumefaiens* include Suraninpong et al. (2004), Mahalakshmi et al. (2006), Yadav et al. (2012) and Bhajan et al. (2019) with up to 90% transformation frequency and positive identification of transgenic plants, especially from T_0 to T_2.

9.3 SUMMARY

Legume species in general are considered to be highly recalcitrant to genetic manipulation, especially using *Agrobacterium tumefaciens* compared to other genomic tools. Although transgenic shoot regeneration using this technique has emerged as the most promising and affordable tool compared to techniques such as electroporation and microprojectile bombardment, the employment of these tools for efficient, stable and rapid genetic improvement in legumes is still inefficient and somewhat unachievable. So far, transgenic mung bean plants produced through *Agrobacterium*-mediated genetic transformation are still the most efficiently and reproducibly developed using cotyledonary node system among all legume crops, including soybean. But like most crops, mung bean also suffers from elevated susceptibility to biotic stresses, including diseases caused by fungi, bacteria, viruses and attack by insect pests. For the role of these factors in growth and productivity inhibition in soybean, see the next chapter (Chapter 10). Mung bean transformation is now comparable with *in vitro* regeneration and *Agrobacterium*-mediated genetic transformation of several cereal crops that have been found to be amenable to rapid and efficient recovery of transgenic plants.

Another advantage is that uniformly produced year-round explant sources can be readily found for these crops. Furthermore, most legume plant transformations use cotyledonary nodes or axillary bud regions as explants to effect incorporation of the transgenes which are normally identified using binary plasmids carrying *GUS* and *hpt*/*nptII* marker genes. Often these selectable marker genes are required for the initial selection and identification of transformed shoots. Stable integration of the marker genes and transgene, including its inheritance in subsequent generations may be efficiently confirmed through molecular analysis. PCR for positive identification of the *nptII* gene and Southern blotting or electrophoresis to confirm putative transformation have also been commonly used in all legume and non-legume crops (Mahalakshmi et al. 2006; Yadav et al. 2012). Based on advancements so far made, there is no doubt that gene transfer (whether direct or indirect) has opened new ways for the incorporation of exogenous DNA molecules between relative species or genotypes that are not genetically compatible.

All of these modern technologies, including gene editing, are now used to complement conventional breeding that is highly required for agricultural expansion. Nutritional and health benefits associated with each and every legume crop species covered above clearly indicate that agricultural expansion is necessary to support the environment, society, economy and even to inform the developments of policies (settlements, regulations, laws, etc.). This is so because all humans and animals depend on agriculture for food, shelter, generating income and providing health-promoting benefits. Therefore, the breeding of legume crops contributes immensely to sustainable agriculture. Generally, *Agrobacterium*-mediated genetic transformation of legumes offers the following to agriculture and society at large: (1) increasing crop productivity by enhancing physical yield, quality of grains and establishing pest and disease resistance, (2) improving resource use efficiency through the optimisation of land, water, fertilisers and other chemical inputs and (3) reducing negative environmental effects such as pollution and the emission of greenhouse gases.

9.4 ABBREVIATIONS

ALS Acetolactate synthase
AB Ascochyta blight
BA Benzyladenine
BM Bari Masur
BGM *Botrytis* grey mould
Bt *Bacillus thuringiensis*
DNA Deoxyribonucleic acid
GUS *β-glucuronidase*
MS Murashige and Skoog
PCR Polymerase chain reaction

REFERENCES

Abudulai M, Seini SS, Haruma M, Mohammed AM and Asante SK. (2016). Farmer participatory pest management evaluations and variety selection in diagnostic famer field for a in cowpea in Ghana. *African Journal of Agricultural Research* 11, 1765–1771.

Aggarwal PR, Nag P, Choudhary P, Chakraborty N and Chakraborty S. (2018). Genotype-dependent Agrobacterium rhizogenes-mediated root transformation of chickpea: A rapid and efficient method for reverse genetics studies. *Plant Methods* 14(55), 1–13.

Akcay VC, Mahmoudian M, Kamci H, Yacel M and Oktem HA. (2009). *Agrobacterium tumefaciens*-mediated genetic transformation of a recalcitrant grain legume, Lentil (*Lens culinaris* Medik). *Plant Cell Report* 28, 407–417.

Amugune NO, Anyango B and Mukiama TK. (2011). *Agrobacterium*-mediated transformation of common bean. *African Crop Science Journal* 19(3), 137–147.

Barampama Z and Simard RE. (1993). Nutrient composition, protein quality and antinutritional factors of some varieties of dry beans (*Phaseolus vulgaris*) grown in Burundi. *Food Chemistry* 47(2), 159–167.

Bett B, Gollasch S, Moore A, Harding R and Higgins TUV. (2019). An improved transformation system for cowpea (*Vigna unguiculata* L. Walp) via sonication and a kanamycin-geneticin selection regime. *Frontiers in Plant Science* 10(219), 1–10.

Bhajan SK, Begum S, Islam MN, Hoque MI and Sarker RH. (2019). *In vitro* regeneration and *Agrobacterium*-mediated genetic transformation of local varieties of mung bean (*Vigna radiata* (L.) Wilczek). *Plant Tissue Culture and Biotechnology* 29(1), 81–97.

Boettinger P, Steinmetz A, Schieder O and Pickardt T. (2001). *Agrobacterium*-mediated transformation of *Vicia faba*. *Molecular Breeding* 8(2), 243–254.

Bogale GA, Maja MM and Gebreyohannes GH. (2021). Modelling the impacts of climate change on faba bean (*Vicia faba* L.) production in Welmera area, Central Ethiopia. *Heliyon* 7(10), e08176, 1–10.

Castro-Guerrero NA, Isidra-Arellano M, Mendoza-Coatl DG and Valdes-Lopez O. (2016). Common bean: A legume model on the rise for unravelling responses and adaptations to iron, zinc and phosphate deficiencies. *Frontiers in Plant Science* 7, Article No. 600, 1–7.

Das SK, Shethi KJ, Hoque MI and Sarkar RH. (2012). *Agrobacterium*-mediated genetic transformation in lentil (*Lens culinaris* Medik) followed by *in vitro* flower and seed formation. *Plant Tissue Culture and Biotechnology* 22(1), 13–26.

Das SK, Shethi KJ, Hoque MI and Sarkar RH. (2019). *Agrobacterium*-mediated genetic transformation of lentil (*Lens culinaris* Medik) with chitinase gene followed by *in vitro* flower and pod formation. *Plant Tissue Culture and Biotechnology* 29(1), 99–109.

De Khathen A and Joconsen H. (1990). *Agrobacterium tumefaciens*-mediated transformation of *Pisum sativum* L. using binary and cointegrate vectors. *Plant Cell Reports* 9(5), 276–279.

Gulati A, Schrger P and McHughen A. (2002). Production of fertile transgenic lentil (Lens culinaris Medix) plants using particle bombardment. *In Vitro Cell and Developmental Biology–Plant* 38, 316–324.

Hanafy M, Pickardt T, Kiesecker H and Jocobsen HJ. (2005). *Agrobacterium*-mediated transformation of faba bean (*Vicia faba* L.) using embryo axes. *Euphytica* 142, 227–236.

Hou D, Yousaf L, Xue Y, Hu J, Wu J, Hu X, Feng N and Shen Q. (2019). Mung bean (*Vigna radiata* L.): Bioactive polyphenols, polysaccharides, peptides and health benefits. *Nutrients* 11(6)(1238), 1–28.

Jaiwal PK, Kumari A, Ignacimuthu S, Potrykus I and Sautter C. (2001). *Agrobacterium tumefaciens*-mediated genetic transformation of mung bean (*Vigna radiata* L. Wilczek)- a recalcitrant grain legume. *Plant Science* 161(2), 239–247.

Janusauskaite D and Razbadauskiene K. (2021). Comparison of productivity and physiological traits of faba bean (*Vicia faba* L.) varieties under conditions of boreal climatic zone. *Agronomy* 11(707), 1–16.

Karmakar S, Molla KA, Gayen D, Karmakar A, Das K, Sarkar SN, Datta K and Datta SK. (2019). Development of a rapid and highly efficient *Agrobacterium*-mediated transformation system for pigeon pea [*Cajanus cajan* (L.) Millsp.]. *GM Crops and Food* 10, 115–138.

Kyei-Boahen S, Savala CEN, Chikoye D and Abaidoo R. (2017). Growth and yield response of cowpea to inoculation and phosphorus fertilisation in different environments. *Frontiers in Plant Science* 8(646), 1–13.

Lavin M, Pennington RT, Klitgaard BB, Sprent JI, de Lima HC and Gasson PE. (2001). The dalbergioid legumes (Fabaceae): delimitation of a pantropical monophyletic clade. *American Journal of Botany* 88(3), 503–533.

Mahalakshmi LS, Leela T, Kumar SM, Kumar BK, Naresh B and Devi P. (2006). Enhanced genetic transformation efficiency of mung bean by use of primary leaf explants. *Current Science* 91(1), 93–99.

Mangena P. (2020). Breeding of legumes for stress resistance. In Mangena P (eds), In Advances in Legume Research: Physiological Responses and Genetic Improvement for Stress Resistance, Vol 1. Bentham Science, Singapore, Singapore SG. pp. 1–16.

Montoya CA, Lalles JP, Beebe S and Leterme P. (2010). Phaseolin diversity as a possible strategy to improve the nutritional value of common beans (*Phaseolus vulgaris*). *Food Research International* 43, 443–449.

Nair RM, Pandey AK, War AR, Hanumantharao B, Shwe T, Alam AKMM, Pratap A, Malik SR, Karimi R, Mbeyagala EK, Douglas CA, Rane J and Schafleitner R. (2019). Biotic and abiotic constraints in mung bean production- progress in genetic improvement. *Frontier in Plant Science* 10(1340), 1–24.

Puonti-Kaerlas J, Erikson T and Engstrom P. (1990). Production of transgenic pea (*Pisum sativum* L.) plants by *Agrobacterium tumefaciens*-mediated gene transfer. *Theoretical and Applied Genetics* 80(2), 246–252.

Sabbah M, Giosafatto CVL, Esposito M, Pierro PD, Mariniello L and Porta R. (2019). Transglutaminase cross-linked edible films and coatings for food applications. In Kuddus M (ed), *Enzymes in Food Biotechnology: Production, Applications and Future Prospects*. Academic Press, London Wall. pp. 369–388.

Sanyal I, Singh AK, Kaushik M and Amla DV. (2005). *Agrobacterium*-mediated transformation of chickpea (*Cicer arietinum* L.) with *Bacillus thuringiensis* cry1Ac gene for resistance against pod borer insect *Helicoverpa armigera*. *Plant Science* 168(4), 1135–1146.

Sathe SK. (2016). Beans: Overview. In Wrigley C, Seetharaman K, Corke H and Faubion J (eds), In *Encyclopedia of Food Grains*, 2nd ed. Academic Press, Boston, Massachussetts. pp. 297–306.

Sharma KK, Bhatnagar-Mathur P and Jayanand B. (2006). Chickpea (*Cicer arietinum* L.). In Wang K (ed), *Agrobacterium Protocols*, Vol 1, 2nd ed. Humana Press, Totowa, New Jersey. pp. 313–323.

Sheahan M and Barrett CB. (2017). Ten striking facts about agricultural input use in sub-Saharan Africa. *Food Policy* 67, 12–25.

Solis-Ramos LY, Ortiz-Pavon JC, Andrade-Torres A, Porras-Murillo R, Angulo AB and de la Serna EC. (2019). *Agrobacterium tumefaciens*-mediated transformation of common bean (*Phaseolus vulgaris*) var. Brunca. *Revista de Biologia Tropical* 67(2), 1–15.

Srivastava J, Datta S and Mishra SP. (2017). Development of an efficient *Agrobacterium* mediated transformation system for chickpea (*Cicer arietinum*). *Biologia* 72(2), 153–160.

Suraninpong P, Chanprame S, Cho H-J, Widholm JM and Waranyuwat A. (2004). *Agrobacterium*-mediated transformation of mungbean [Vigna radiata (L.) Wilczek]. *Walailak Journal of Science and Technology* 1(2), 38–48.

Surekha C, Beena MR, Arundhati A, Singh PK, Tuli R, Datta-Gupta A and Kirti PB. (2005). *Agrobacterium*-mediated genetic transformation of pigeon pea (*Cajanus cajan* (L.) Millsp) using embryonal segments and development of transgenic plants for resistance against Spodoptera. *Plant Science* 169(6), 1074–1080.

Svabova L, Smykal P, Griga M and Ondrej V. (2005). *Agrobacterium*-mediated transformation of *Pisum sativum in vitro* and *in vivo*. *Biologia Plantarum* 49(3), 361–370.

Tang D, Dong Y, Ren D, Li L and He C. (2014). A review of phytochemistry, metabolite changes and medicinal uses of the common food mung bean and its sprouts (*Vigna radiata*). *Chemistry Central Journal* 8(4), 1–9.

Verma AK and Chand L. (2005). *Agrobacterium*-mediated transformation of pigeon pea (*Cajanus cajan* L.) with *uidA* and *cry1A(b)* genes. *Physiology and Molecular Biology of Plants* 11(1), 99–109.

Yadav SK, Katikala S, Yellisetty V, Kannepalle A, Narayana JL, Maddi V, Mandapaka M, Shanker AK, Bandi V and Bharadwaja KP. (2012). Optimisation of *Agrobacterium* mediated genetic transformation of cotyledonary node explants of *Vigna radiata*. *Springer Plus* 1(59), 1–8.

10 Transgenic Plants for Biotic Stress Resistance in Soybean

10.1 INTRODUCTION

Soybean constitutes a large number of varieties that undergo continuous genetic manipulations to enhance their useful genomic and functional growth properties. Successful breeding of stress-tolerant varieties for cultivation under different farming systems may result in reduced production costs, limited use of agrochemicals and increased yields. Those are some of the advances achieved through biotechnology other than conventional breeding, which is rather limited in terms of circumventing sexual barriers across species. In general, legume crops having pathogen and herbicide resistance exhibit better growth and yield characteristics, with strong competitiveness against weeds and other biological pests. As established several decades ago, the revolution in genomic research led to the development of many sophisticated and advanced crop improvement techniques that can be applied across a whole range of species such as cowpea, faba bean, lentil, mung bean, pea and soybean, etc.

However, interest in indirect genetic engineering, chemical- or physical-based mutation breeding, marker-assisted selection, quantitative trait loci and genome editing have expanded research beyond biotic stress resistance. These techniques play a critically important role in applications such as the manufacturing of bioenergy, crop engineering for the expression of valuable bioactive compounds and recombinant proteins. This chapter briefly reviews advances made in conferring resistance of some soybean varieties to diverse biotic stress factors (bacteria, fungi, insects, parasitic nematodes and viruses), and also looks into possible ways in which these stress factors can be managed and eradicated using *Agrobacterium tumefaciens*-mediated genetic transformation. This review shows how biotechnological tools such as this one could be used to provide beneficial functions in pest management through genetic, physiological and morphological improvements, especially when coupled with other farming practices.

10.2 DEFINING BIOTIC STRESS

Biotic stress can be broadly defined as any living component, stress-inducing factor or stress constraint that prevents the plant from achieving its full genetic potential. Authors such as Lincoln Taiz and Eduardo Zaiger referred to this kind of stress,

DOI: 10.1201/b22829-10

including abiotic stress, as growth-inhibiting conditions that may not allow plants to achieve maximum growth and reproductive capacity (Taiz et al. 2015). However, this phenomenon can be easily understood by quantifying vegetative and yield characteristics, such as plant height, leaf/branch number, biomass, fruit and seed quantities. These attributes can be reliably and definitively assessed following plant exposure and attack by biotic stress factors. Therefore, under growth-limiting conditions, biotic stress refers to all negative influences caused by living organisms such as parasitic nematodes, viruses, disease-causing bacteria, fungi, arachnids, weeds and insect pests. Some common microbial and insect pests that cause damage and diseases in legumes in general, including soybean and other crops, are summarised in Table 10.1.

The table indicates some of the most common types of living organisms that co-exist with plants in their immediate environment. Although some of these organisms have mutually beneficial interactions with plants, other parasitic or pathogenic species can become detrimental. These organisms include microbial pathogens like *Xanthomonas campestris pv. phaseoli*, *Fusarium oxysporum f.sp. ciceris*, *Leveillula taurica cv. Arn*, Alfalfa mosaic virus (AMV) and herbivorous insects like leafhoppers as well as beetles (Table 10.1). In response to dealing with biotic stress, plants have evolved intricate defence mechanisms to deal with the harmful effects of pests and microbial pathogens. These involve physiological mechanisms that plants induce in order to cope and deal with the continuous attacks that often lead to extreme losses in crop productivity. Accordingly, defence mechanisms are usually triggered either when the toxic secondary metabolite released by the pathogen reaches the plant's internal cellular compartments or the system is activated upon immediate detection of the attack through inducible defence which utilises specific detection and signal transduction pathways (Iqbal et al. 2021).

Both specific detection system and signal transduction can sense the presence of an herbivore or pathogen and then alter its gene expression and metabolism accordingly to counter the stress. Plant normally use mechanical barriers such as the cell wall (with silica in certain species), cuticle (a waxy outer layer), periderm or papillae formation as a constitutive defence system. The complex structure of papilla cells is formed between the cells' plasma membrane and the inside of the cell wall. Huckelhoven (2014) referred to these cell appositions as the ones responsible for preventing fungal attempts to penetrate the cell walls of plants. This report also revealed that the molecular composition of these papillae differs significantly from those of the primary and secondary cell walls. Plants also use toxic secondary metabolites to defend against insects and other herbivores. However, biotic stresses reduce growth rates and cause major losses pre- and post-harvesting, where climate plays a key role in determining the type of stress factors that are imposed on crop plants, as well as the plants' ability to resist such attacks. Furthermore, the stress negatively influences the rate of photosynthesis since insect pests reduce leaf area by chewing the leaves.

Meanwhile, microbial pathogens such as *Xanthomonas axonopodis* pv. *citri* also reduce photosynthesis by negatively affecting the activity of key enzymatic proteins such as ribulose 1,5 bisphosphate carboxylase (Rubisco), Rubisco activase and adenosine triphosphate synthase (ATPase) (Garavaglia et al. 2020).

TABLE 10.1
Some of the most common biotic stress factors negatively affecting legume crops under diverse environmental conditions

Category	Species (disease/common name)
Bacteria	*Pseudomonas syringae pv. phaseolicola* (halo blight)
	Pseudomonas syringae pv. syringae (bacterial brown spot)
	Xanthomonas campestris pv. phaseoli (bacterial blight)
Fungi	*Fusarium oxysporum f.sp. ciceris* (*Fusarium* wilt)
	Fusarium solani (black root rot)
	Leveillula taurica cv. Arn (powdery mildew)
	Erysiphe spp. (powdery mildew)
	Uromyces cicer-arietini [Gorgn.] (rust)
	Rhizoctonia spp. (dry/wet root rots)
	Sclerotium rolfsii (collar rot)
Nematodes	*Meloidogyne* spp. (root knot)
Viruses	*Alfalfa mosaic virus (AMV)*
	Beet Western yellow virus (BWYV)
	Broad bean mosaic virus (BBMV)
	Seed borne mosaic virus (SBMV)
	Broad bean wilt virus (BBWV)
	Bean golden mosaic virus (BGMV)
Insects pests	*Empoasca* spp. (leafhopper)
	Aphis craccivora (aphid)
	Ophiomyia phaseoli syn. Melanagromyza phaseoli (bean fly)
	Ootheca mutabilis (beetle)
	Mylabris spp. (beetle)
	Medythia guaterna (beetle)
	Nezara spp. (bug)
	Anoplocnemis spp. (bug)
	Riptortus spp. (bug)
	Acanthomia spp. (bug)

References: bacteria (Schwartz 2011); fungi (Perfect et al. 1999); nematodes (Davis and Mitchum 2005); viruses (Chatzivassiliou 2021); insects pests (Edwards and Singh 2006; Singh and van Emden 1979).

Further below, a literature survey and discussion of the existing wide range of biotic stress factors and the extent to which these organisms affect soybean growth, yield and beyond are given. Highlights on the diverse biotechnological mechanisms developed globally to help crop plants in overcoming biotic stress are also provided. Then, this chapter, at a more advanced point, elaborates on constitutive and inducible defences, briefly taking note of the beneficial interactions that exist amongst plants and microorganisms, and describes the role that other ecological factors (for example, climatic factors, physiographic factors and animals) play in biotic stress evolution.

10.3 PLANT-PATHOGEN INTERACTIONS

Soybeans like many other crop plants have naturally evolved defence mechanisms in response to herbivores, environmental stress constraints and pathogens. Defence mechanisms in these plants consist of physical and chemical barriers, in addition to active defence responses induced after the perception of potential stress or invasion by microbial pathogens. Nevertheless, there are emerging challenges relating to the loss of resistance due to decreased genetic variations, especially the narrow diversity in modern cultivars. Soybeans that are currently cultivated for agricultural purposes are mainly developed from the already cultivated gene pool. This is where losses of resistance-based genetic diversity are experienced. For instance, soybean-cyst nematode (SCN) *Heterodera glycine Ichinohe* is considered as one of the most damaging soybean pests across the globe. There are indications that the current soybean gene pool is losing resistance against SCN (Kofsky et al. 2021).

Soybean cultivars show tolerance to SCN such as Peking, which is the main source of resistance genes to race 5 soybean lines in the United States. This gene exhibits polyamine synthesis to promote the structural integrity of root cell walls. Using naturally available resistance capacity to improve and develop new soybean varieties will help maintain food security without compromising the environment. While the existing interplay between biotic and abiotic stresses against plants still remains elusive and unpreventable. Plants must also defend themselves against pathogen invasion by creating a physical barrier inhibiting penetration by microbial pathogens, and induce programmed cell death (PCD) to prevent the spread of the disease to uninfected cells. In fact, gene networks controlling defence responses must be identified and characterised in order to confer durable broad-spectrum resistance against pathogens in soybean (Liu et al. 2015).

Functional genomic studies are necessary since the natural active defence responses currently used by plants may no longer be adequate. Both structural modifications and biochemical response reactions might not be effective in the defence of the plant. This may be so because the average estimated annual soybean yield losses owing to diseases worldwide continue to increase. To date, the use of disease-resistant cultivars developed through techniques such as *Agrobacterium tumefaciens*-mediated indirect gene transfer has been the major method for controlling most soybean pathogens. Plant interactions with pathogens still offer an avenue for improvement since this association may not be compatible resulting in plant disease resistance. When the interaction between plants and pathogens is compatible, then disease symptoms will be formed. This is similar to avirulent pathogens which exist that may not be able to successfully infect and invade their hosts. Such pathogens will fail to induce disease symptoms and as a result lead to an incompatible interaction between the plant and the pathogen.

10.4 SOYBEAN-PATHOGEN SPECIFICITY GENES

Plant pathogens occupy a broad range of environmental conditions, particularly the inhabitable ecological niche on earth. Most of these pathogenic species live on their host plants and derive their nutrients from their host often becoming completely parasitic. Frequently, fungal pathogens have demonstrated this kind of habituation accompanied by the competency to expand their host range and cause opportunistic infections (Mangena and Mkhize 2021). However, the ability for a pathogen to cause a disease in plants or a plant to be resistant to one or several pathogens at the same time is a phenomenon controlled and determined by certain genomic sequences. Host resistance, for example, is controlled by one or more resistance genes. The presence of these genes is thought to be a key factor in preventing the infection and the disease progression. Similarly, the occurrence of genes required by the pathogen for pathogenicity, specificity and virulence against plant hosts also plays a critical role in ensuring a successful infection and disease development.

These genetic sequences enable pathogens to serve as single or broad-host range disease-inducing factors causing symptoms in a large number of different host organisms. But host resistance to diseases requires pathogen recognition where both specific avirulence genes and resistance genes have to be present in the pathogen and host, respectively, in order for the host to develop resistance. In soybean, the causal agent of bacterial blight (*Pseudomonas syringae* pv. glycinea) has been the first one demonstrating this gene-to-gene specific interaction (Huynh et al. 1989). Usually, resistant soybean varieties typically develop a hypersensitive reaction (HR) by collapsing the mesophyll cytoplasm of the infected cells to inhibit further infection and formation of bacterial blight. This effect takes place mainly in the site of infection. In another study, it was revealed that soybean plants were susceptible to certain races of the pathogen *Phytophthora megasperma* f.s.p. glycinea (*Pmg*). The reaction indicated that pathogen isolates differed according to virulence and pathogenicity, and the affected soybean lines were also tolerant to a limited number of *Pmg* isolates (Thomison et al. 1988).

Apart from these, soybean is already known for hosting a variety of pathogens that cause significant losses to its yield, including *Phakopsora pachyrhizi*, *Heterodera glycine* and soybean mosaic virus. Pathogens predominantly affecting this crop fall within groups of bacteria, viruses, oomycetes, fungi and nematodes. Organisms from these groups possess conserved unique features that can be targeted to develop plant resistance through modern plant breeding programmes (Whitham et al. 2016). The understanding and analysis of soybean-pathogen interactions together with their mechanism of host or strain specificity mediated by various gene sequences continue to provide insights that underline pathogenicity and plant immunity. This information is necessary in dealing with constraints that negatively impact on soybean production by utilising genetic resources that show resistance to biotic stresses. Additionally, emerging pathogens such as *Kosakonia cowannii* form part of the new pathogens affecting soybeans. This bacterium causes disease symptoms resembling those of bacterial blight and bacterial pustule. The evolution of novel pathogens in the environment poses serious threats for crops, and serves as a clear indication that genetic engineering protocols must be improved that will effectively deal with future emerging threats.

10.5 SOYBEAN DEFENCE AGAINST BIOTIC STRESS

Epiphytic and endophytic pathogens and insect pests remain abundant in the natural environment. Some of these disease-causing agents exhibit very quick metabolism, as well as the ability to rapidly colonise new ecological niches and hosts. Unfortunately, unlike when dealing with insect attacks, disease-causing pathogens produce symptoms that are not distinguishable. As a result, crops like soybeans are easily attacked by numerous potential disease factors of bacterial or viral origin. Meanwhile, bugs, beetles and other insect pests such as thrips and aphids are easily identifiable. According to Krawczyk and Borodynko-Filas (2020) diseases such as bacterial blight are also easily recognised, together with bacterial pustule. Bacterial pustule causes premature defoliation and results in reduced seed sizes as well as poor grain quality.

On the other hand, bacterial blights can be transmitted by seeds, occurring early during the cultivation seasons and causing water-soaked spots that turn yellow or brown when leaf tissues die. Yellow and brown spots merge to form large dead patches on the leaves of infected soybean plants. Bacterial blight can also occur on infected pods, seeds, stems and petioles (Chen et al. 2021). However, as previously indicated, soybean plants may be able to defend themselves against these microbials if they contain resistant (*R*) genes in their genomes, but with the infecting pathogen carrying an avirulence gene. Experimental evidence conclusively suggested that most pathogens contain avirulence genes whose products are involved in determining host specificity (Whitham et al. 2016). For instance, *P. syringae* encode gene proteins that are injected into host cells using a type III secretion system as part of the effector protein cocktail that controls and is intended to suppress the host's immune system. Primary infection usually starts with the growth of the pathogen on the plant surface without invading internal tissues.

The pathogen typically waits for conditions to be favourable before starting the infection process. In this case, the plant defence structures such as waxes, cuticle and cell wall will be used to avoid and prevent the infection, assisted by other morphological features such as leaf shape, folding, venation, trichomes, stomata and the presence of lenticels. Secondary defence will then constitute a chemical defence through the production of secondary metabolites which are toxic to the pathogens and insect pests. Recently, genomic sequencing revealed hundreds of resistance genes clustered on chromosomes known as disease resistance quantitative trait loci (QTL). In soybean cultivar Williams 82 it was found that 319 nucleotide binding site-leucine rich repeat (NBS-LRR) genes form the most predominant type of resistance. NBS-LRR genes underlying QTL in soybeans control disease resistance traits and are responsible for encoding proteins that prevent attack by insect pests or eliminate pathogenic microbes.

Once such proteins are induced, like the cytoplasmic NBS-LRR proteins Rpg1b and Rpg1r, their expression activates a number of physiological defence response reactions, including PCD and the production of antimicrobial secondary metabolites known as phytoalexins in the site of infection as well as the surrounding tissues. Research on soybean-pathogen interactions has led to the development of vector systems like *Agrobacterium tumefaciens*-mediated genetic transformation for the transfer and incorporation of foreign specific resistance genes into host plants. Plant transformation efforts must be intensified, coupled with functional genomics to confer resistance by cloning *R* genes from other sources.

10.6 TRANSGENICS CONFERRING BIOTIC STRESS RESISTANCE

Pathogen-related (PR) proteins are expressed and accumulated in both infected and uninfected host tissues during plant defence responses. The induction of these specific proteins only during invasion suggests that they play an important role in defending the plants against microbial and insect attackers. Although plants naturally and inherently possess the ability to induce PR proteins, together with other secondary metabolites, genetic manipulation of crops is still required to improve resistance in genotypes that are less tolerant. Both recombinant DNA technology and modern breeding practices of soybean cultivars involving the strong selection or introgression of such agronomic traits must be prioritised. For the past three decades, a large number of transgenic soybean varieties have been developed for biotic stress resistance and more are still underway. Among some of the incorporated genes, these transgenic plants express proteins such as cystatins, trypsin inhibitors, protease inhibitors and synthetic crystal gene protein products.

Soybean plants derived from somatic embryos of cultivar Jack were protected from damage from corn earworm (*Helicoverpa zea*), soybean looper (*Pseudoplasia includens*), tobacco budworm (Heliothis virescens) and velvetbean caterpillar (Anticarsia gemmatalis) after being genetically transformed. Three lines of soybean were recovered following genetic transformation using microprojectile bombardment with a synthetic *Bacillus thuringiensis* insecticidal crystal protein gene (*BtCryIAc*). Furthermore, all plants were found to be fertile, with varying resistance levels ranging from 3 to 40% (Stewart et al. 1996). In another study, *Agrobacterium tumefaciens* strain LBA4404 was used to establish transgenic soybean plants expressing the *cry8*-like gene conferring resistance to *Holotrichia parallela*, also known as the dark black chafer. The *cry8*-like gene for *Bt* strain HBF-18 was introduced into soybean cultivar Jinong 28 to develop a promising strategy for engineering pest tolerance (Qin et al. 2019).

However, a thorough assessment of recent and current literature demonstrates that many transgenic crops containing *Bt* toxin (cryIAb protein) are being introduced. This is taking place despite suggestions made by various reports that *Bt* crops have detrimental effects on other living organisms such as the Lepidopteran monarch butterfly (*Danaus plexippus*) and the environment in general (Then and Bauer-Panskus 2017). Protease inhibitors were also reported to be a promising complement for *Bt* toxins in the development of insect-resistant transgenic soybeans. However, challenges such as limited specificity against proteolytic enzymes and ubiquity of metabolic processes that are dependent on protease enzymes still raise serious questions about their efficacious applications in genetic transformation (Schluter et al. 2010).

Furthermore, in soybean, endogenous protease inhibitors such as cysteine proteases regulate several processes that contribute immensely to the crop's tolerance to abiotic stress factors. Although these proteins have many functions that remain poorly characterised, particularly under biotic and abiotic stress, valuable information on their localisation, regulation, target organs and tissues used for expression was succinctly reviewed by Mangena (2020a). The role of these papain-like cysteine proteases and their inhibitors in soybean plants against abiotic stress will be comprehensively discussed in the next chapter.

10.7 SUMMARY

The application of plant genetic transformation for the development of transgenic soybean plants with tolerance to the various biotic stress factors is still very stagnant. Some of the major reasons attributed to these gradual advancements involve the lack of abundant specific *R* genes, requirement for the presence of avirulence genes and species incompatibility and absence of exogenous genetic resources for the development of newly improved lines. Genetic engineering of resistant plants is an important measure that can be used widely to address current issues facing the human race. Such issues include the depletion of biodiversity, climate change, deforestation, drought, food shortages, poverty, unemployment, water and soil pollution. Insect- and disease-resistant soybean plants will ensure that the use of agrochemicals, disease spreads, yield losses and environmental pollution are minimised since the cultivation of transgenic plants requires low agricultural inputs.

Naturally, all plants, including soybeans that are not genetically engineered, mainly rely on both morphological and physiological defence mechanisms to protect themselves from attacks by microbial and insect pathogens. As plant diseases continue to significantly decrease crop production worldwide, integrating crop rotation and reduced use of pesticide with highly enhanced resistant varieties is a major prerequisite. Plant transformation serves as one of the most important modern breeding tools that may successfully complement conventional breeding which has a limited scope and remains an extremely time-consuming process. Currently, more than one-third of the world's edible oils and two-thirds of protein meals are derived from genetically engineered soybeans (Mangena 2020b). These achievements were made amid the increased host of technology concerns and negative perception of plant transformation. However, the continuing loss of resistance against diseases and pests by the presently cultivated soybean varieties only reiterates the call to optimise and establish new *in vitro* or *in-planta*-based genetic manipulation protocols.

10.8 ABBREVIATIONS

AMV Alfalfa mosaic virus
ATPase Adenosine triphosphate synthase

Bt	Bacillus thuringiensis
Cry	Crystal protein gene
DNA	Deoxyribonucleic acid
NBS-LRR	Nucleotide binding site-leucine rich repeat
PCD	Programmed cell death
PR	Pathogen-related proteins
QTL	Quantitative trait loci
R	Resistant gene
SCN	Soybean cyst nematode

REFERENCES

Chatzivassiliou EK. (2021). An annotated list of legume-infecting viruses in the light of metagenomics. *Plants* 10(1413), 1–17.

Chen NWG, Ruh M, Darrasse A, Foucher J, Briand M, Costa J, Studholme JD and Jacques M-A. (2021). Common bacterial blight of bean: A model of seed transmission and pathological convergence. *Molecular Plant Pathology* 22(12), 1462–1480.

Davis EL and Mitchum MG. (2005). Nematodes. Sophisticated parasites of legumes. *Plant Physiol* 137(4), 1182–1188.

Edwards O and Singh KB. (2006). Resistance to insect pests: What do legumes have to offer? *Euphytica* 147, 273–285.

Garavaglia BS, Thomas L, Gottig N, Zimaro T, Garofalo CG, Gehring C and Ottado J. (2020). Shedding light on the role of photosynthesis in pathogen colonization and host defense. *Communicative and Integrative Biology* 3(4), 382–384.

Huckelhoven R. (2014). The effective papilla hypothesis. *New Phytologist* 204, 438–440.

Huynh TV, Dahlbeck D and Staskawicz BJ. (1989). Bacterial blight of soybean: Regulation for a pathogen gene determining host cultivar specificity. *Science* 22(245), 1374–1377.

Iqbal Z, Iqbal MS, Hashem A, Allah EFA and Ansari MI. (2021). Plant defense response to biotic stress and its interplay with fluctuating dark/light conditions. *Frontiers in Plant Sciences* 12(631810), 1–22.

Kofsky J, Zhang H and Song B-H. (2021). Novel resistance strategies to soybean cyst nematode (SCN) in wild soybean. *Scientific Reports* 11(7967), 1–13.

Krawczyk K and Borodynko-Filas N. (2020). Kosakonia cowannii as the new bacterial pathogen affecting soybean (Glycine max Willd.). *European Journal of Plant Pathology* 157, 173–183.

Liu J-Z, Graham MA, Pedley KF and Whitham SA. (2015). Gaining insight into soybean defense responses using functional genomics approaches. *Briefings in Functional Genomics* 14(4), 283–290.

Mangena P. (2020a). Genetic transformation to confer drought stress tolerance in soybean (Glycine max L.). In Guleria P, Kumar V and Litchtfouse E (eds), *Sustainable Agriculture Reviews 45, Legume Agriculture and Biotechnology*, Vol 1. Springer Nature, Switzerland AG. pp. 193–224.

Mangena P. (2020b). Phytocystatins and their potential application in the development of drought tolerance plants in soybean (Glycine max L.). *Protein and Peptide Letters* 27(2), 135–144.

Mangena P and Mkhize P. (2021). Analysis of cross-reactivity, specificity and the use of optimised ELISA for rapid detection of Fusarium spp. In Khan MS (ed), *Frontiers in Molecular Pharming, Frontiers in Protein and Peptide Letters*, Vol 2. Bentham Books, Singapore. pp. 195–225.

Perfect SE, Hughes HB, O'Connell RJ and Green JR. (1999). Colletotrichum: A model genus for studies on pathology and fungal-plant interactions. *Fungal Genetics and Biology* 27(2–3), 186–198.
Qin D, Liu X-Y, Miceli C, Zhang Q and Wang P-W. (2019). Soybean plants expressing the Bacillus thuringiensis cry8-like gene show resistance to Holotrichia parallela. *BMC Biotechnology* 19(66), 1–12.
Schluter U, Benchabane M, Munger A, Kiggundu A, Voster J, Goulet M-C, Clouteir C and Michand D. (2010). Recombinant protease inhibitor for herbivore pest control: A multi-trophic perspective. *Journal of Experimental Botany* 61(15), 4167–4183.
Schwartz HF. (2011). Bacterial diseases of beans. Crop Series Diseases. *Colorado State University Extension* 2(913), 1–3.
Singh SR and Van Emden HF. (1979). Insect pests of grain legumes. *Annual Review of Entomology* 24, 255–278.
Sterwart Jr CN, Adang MJ, All JN, Boerma R, Cardineau G, Tucker D and Parrott WA. (1996). Genetic transformation, recovery and characterisation of fertile soybean transgenic for a synthetic Bacillus thuringiensis cryIAc gene. *Plant Physiology* 112(1), 121–129.
Taiz L, Zeiger E, Moller IM and Murphy M. (2015). *Plant Physiology and Development.* Sinauer Associates, London. pp. 756–60.
Then C and Bauer-Panskus A. (2017). Possible health impacts of *Bt* toxins and residues from spraying with complementary herbicides in genetically engineered soybeans and risk assessment as performed by the European Food Safety Authority EFSA. *Environmental Sciences Europe* 29(1), 1–11.
Thomison PR, Thomas CA, Kenworthy WJ and McIntosh MS. (1988). Evidence of pathogen specificity intolerance of soybean cultivars to *Phytophthora* rot. *Crop Science* 28(4), 714–715.
Whittham SA, Qi M, Innes RW, Ma W, Lopes-Caitar V and Hewezi T. (2016). Molecular soybean-pathogen interactions. *Annual Review of Phytopathology* 54, 443–468.

11 Transgenic Plants for Abiotic Stress Resistance in Soybean

11.1 INTRODUCTION

Recently, crop research has focussed more attention on the development of abiotic stress-tolerant plants, particularly those showing resistance to drought stress. The advancements dramatically changed our understanding of functional and regulatory genomics in relation to abiotic stress, together with disease and insect pest resistance as discussed in the previous chapter (see Chapter 10). Even though the first genetically modified soybean was introduced by Monsanto over two decades ago, a stable transformation for abiotic stress tolerance cannot yet be considered to be a routine. Plant transformation promised from the onset a generation of biotech crops offering protection from losses caused by drought, salinity and heat stress, which are some of the major abiotic stresses. But the commercialisation of transgenic plants since 1996 to date has only presented the production of herbicide-tolerant and insect-resistant soybean cultivars (Roundup Ready (RR) crops by former Monsanto Technology LLC trademarks) (Homrich et al. 2012).

The production of these biotic stress-resistant varieties suggest that the understanding of how genetic engineering works is still a major challenge. It remains a major hurdle after the genomic era, particularly for soybean and other recalcitrant legume crops. There is still an immense failure in revealing how soybean transformation remains genotype-specific, yet the crop is also more vulnerable and susceptible to abiotic stress factors. Since the mechanism and production process of RR-crops is well known, these data, including functional and regulatory information on how plants naturally alter existing pathways by activating stress-response genes, must have already provided a deeper understanding of the physiological and genetic improvement of soybean, especially through *Agrobacterium tumefaciens*-mediated genetic transformation.

High-throughput genetic transformation protocols that facilitate the production of new abiotic stress-tolerant soybean cultivars are required. In addition, the identification and characterisation of molecular markers correlating with the abiotic stress resistance (QTLs) of cultivated genotypes can be expanded to generate more research insights for use in modern breeding tools. In this chapter, major elements of abiotic stress tolerance will be summarised along with the use of genetic transformation utilising *A. tumefaciens* to develop putative soybean plants that efficiently and effectively ameliorate, prevent and repair damages caused by such abiotic stress constraints.

DOI: 10.1201/b22829-11

11.2 DEFINING ABIOTIC STRESS

Living organisms found in specific environments can be negatively affected by non-living factors. This situation reflects and defines what is known as abiotic stress. As a result, an ideal growth environment for a given plant would be the one allowing the plant to achieve its maximum growth and reproductive potential. Soybean plants as one of the living organisms existing and occupying a particular niche in an environment can be adversely affected by abiotic stress factors. These factors include drought, soil salinity, excessive light or heat, heavy metals and chilling stress. All of these environmental factors were found to affect and limit crop productivity across the globe. The use of plant transformation via *Agrobacterium tumefaciens*-mediated genetic transformation may be a promising alternative strategy to overcome the limitations brought by abiotic stress. Achieving this objective could extend the benefits to include the protection of plant health, creating a friendly pollution-free environment, efficient cost-effective breeding programmes and providing food security.

As it is well understood that these factors are hostile to soybean growth and development resulting in greater yield losses, genetic transformation can be utilised to equip crops such as soybean with multi-stress tolerance, relieving them from environmental stress pressures and meeting the demands of the ever-growing global population (He et al. 2018). Unlike animals, plants are generally immobile and as a result, they must always endure pressures imposed by abiotic stress which consistently interfere with vigorous growth. In response to stressful conditions, soybean plants like all other eudicot and monocot plants have evolved elaborate mechanisms used to perceive and respond to environmental stress signals. The variety of mechanisms used by these plants to first sense abiotic stress primarily involve a disruption of their metabolism or transient disturbance to their architectural or structural integrity.

Any disturbances to protein or RNA stability, ion transport through cell walls, decoupling of reactions or other cellular metabolic pathways may cause the soybeans to activate new or already existing stress-response pathways. The induced pathways also involve several stress-inducible gene products enabling the plants to maintain homeostasis during the period of stress (Akpinar et al. 2012). The activation of stress-inducible genes leads to the production of functional proteins providing direct tolerance to abiotic stress, and regulatory proteins responsible for downstream signal transduction coupled with the expression of proteins showing active biological activities against these abiotic stresses. The late embryogenesis abundant proteins (LEA) encoded by a set of genes occur during the last stage of embryonic seed development. These hydrophilic LEA proteins are an example of a response to drought stress signalling that is predominantly taking place in vegetative tissues (Battaglia and Covarrubias 2013).

11.3 ABIOTIC STRESS IMPACT ON SOYBEAN PLANTS

As with every biological system, plant growth and survival depend on a complex network of anabolic and catabolic pathways. It has been clearly observed that the disruption of these metabolic pathways by environmental cues has serious consequences for soybean growth and productivity. About 40% yield losses in major food crops are reported annually. Meanwhile, global figures indicate that over 80% of arable land is prone to drought and 37% of land suffers from elevated levels of salinity stress due to excessive reliance on irrigation practices (Waqas et al. 2019). It has been observed that climate change and global warming result in changes in the patterns of precipitation, extreme weather conditions, accumulation of heavy metals and frequently severe drought stress occurrences. The carbon dioxide (CO_2) emission and global temperature forecasts recently indicated a 2.0–4.9°C (approximately less than 5% warming) increase.

This increase in global temperature is proportional to the anticipated CO_2 emissions and it plays a huge role in the occurrence of abiotic stress and its severity (Raftery et al. 2017). Greenhouse gas emissions do not only cause global warming which results in severe weather patterns but also create air, water and soil pollution, presenting serious risks to animal, human and plant health. The contaminating release of gases, chemicals and heavy metals as industrial wastes or farm inputs also continues to limit crop productivity. While it is critically important to investigate genetic methods of improving plant performance and economic returns to agribusiness, it is also equally important to understand the impact of these abiotic stress factors on growth, development and reproduction in crop plant species as discussed below.

11.3.1 Salinity

Salinity is increasingly becoming a major abiotic stress for soybean, other pulses and grains or cereal crops. In soybean, salinity stress mostly reduces plant height, total biomass and yield production. Soybean has a low tolerance to salt stress, with a threshold of 0.5 $S.m^{-1}$ measured in terms of electrical conductivity. This crop's response to salinity often varies according to precipitation since soluble salts are lowered most of the time by rainfall. Salt concentrations suddenly increases in the topsoil depending on the amount of moisture available in the soil. This causes what are commonly known as salinity peaks which usually take place during summer and winter when waterlogging and drought are likely to occur, respectively (Bustingorri and Lavado 2011). In plants, salinity generally creates nonspecific osmotic stress effects that result in water deficits.

The lack of water in the cells leads to the accumulation of Na^+ and CL^- ions which may increase to more than 100 mM and become cytotoxic. The high concentrations of salts may destabilise cell membranes by denaturing membrane protein structures through dehydration. Na^+ which are more potent denaturants also compete with transport proteins responsible for high affinity uptake and translation of essential macronutrients. Salinity stress is also implicated in the generation of ROS, including singlet oxygen, hydrogen peroxide, hydroxyl radicals and superoxide anions (Mansour 2013). ROS cause damage to plants by attacking macromolecules such as proteins and nucleic acids and disrupt cellular membranes among others.

11.3.2 Heat Stress

Apart from the presence of excessive concentrations of soluble salts in the soil and tissues that suppress plant growth and yield, heat stress has been identified as one of the most damaging environmental factors limiting crop productivity in soybean. High temperatures cause plant stress by disrupting the metabolism through differential effects on protein and enzyme stability. This stress also causes the uncoupling of metabolic reactions and the accumulation of ROS as well as other toxic metabolites (Zhang et al. 2021). Normally, both the increase in temperature and osmotic potential in the cells cause fluctuation in the fluidity of plant cell membranes. The change in membrane fluidity may also signal and activate regulatory mechanisms leading to acclimation (Los and Murata 2004). Heat stress disrupts the transcription and translation of ribonucleic acids (RNAs) and proteins by melting the secondary structures of these molecules.

In addition, this stress can cause a build-up of protein aggregates that interferes with normal cellular functions, especially the functioning of the cytoskeleton and associated organelles. Although cooler temperatures and excessive soil moisture may lead to slow plant growth and increased disease spreads, without an ideal constant temperature of ±29°C soybean stands will be compromised. Vegetatively, heat stress may result in reduced rates of photosynthesis due to the cloning of stomata and limited CO_2 uptake. Poor photosynthetic rates directly impact negatively on vegetative growth, whereby the plants will then accelerate reaching maturity as both temperature and photoperiod play a major role during flowering in soybean (Taiz et al. 2015). Lastly, during reproduction, heat stress causes abortion of flowers, pods and seeds. Leaf loss can also take place, especially under severe heat conditions. This stress negatively impacts soybean growth and productivity regardless of the genotype or determinate or indeterminate growth habit.

11.3.3 Chilling Stress

As indicated in the previous section, heat stress increases membrane fluidity causing denaturation of protein complexes, disruption of electron flow, photosynthesis, homeostasis and regulatory proteins. Metabolic disruptions also take place when plants are exposed to chilling stress, which in contrast results in decreased membrane fluidity. Like all plants, soybeans subjected to lower freezing temperatures must endure the formation of ice crystals. Solid ice may occur outside or inside the cells of the plant. Intracellular freezing always proved to be potentially lethal. This problem is also encountered during artificial exposure of cells to freezing for conservation purposes. One such observation was made in the cryopreservation of *Acer saccharinum* embryonic axes (Wesley-Smith et al. 2014). Cryopreservation refers to a process whereby plant cells, tissues or organs are preserved at temperatures below –150°C for achieving long-term viability and storage.

Meanwhile, extracellular ice formation is nearly always innocuous to the cells. This shows that the damage caused by freezing is a consequence of ice crystal formation in the apoplast and vascular tissues, particularly xylem. The damage is usually caused when unfrozen water within xylem tracheids and vessels or anywhere in the cells normally moves towards the initial crystal ice formation. During this process, the apoplast experiences a negative water potential which eventually becomes lower than that of the symplast (Taiz et al. 2015). However, it was earlier experimentally confirmed that extracellular ice crystals become more damaging by reducing the amount of unfrozen water inside the cells. The evidence provided suggested that the rising solute concentration and diminishing liquid volume, especially water potential (Ψ), induce serious mechanical damage on the cells (Pegg 1987). These findings clearly illustrate that freezing conditions or chilling stress has much in common with drought stress as it is understood in the below discussion.

11.3.4 Drought Stress

Plant cell dehydration caused by the exposure of plants to drought stress forms the most damaging effects compared to any other type of stress. Water makes up a larger proportion of volume in the cells of all living organisms, especially in plants and animals. Perhaps one of the reasons is that water is essential for all growth and physiological processes that are taking place in the cells. As such, a continuously long period of plant growth without precipitation will result in the formation of water stress or drought stress. Water deficit affects plants in every stage of their life cycle, from vegetative to reproductive growth. Soybeans that undergo water deficit will experience dehydration which then affects many basic physiological processes, including major processes such as photosynthesis and respiration (Taiz et al. 2015). Many research findings have showed that soybeans are highly sensitive to drought stress, and this was easily demonstrated by changes in root and shoot biomass.

The failure of roots to absorb sufficient water has direct implications for all metabolic processes supporting root and shoot growth (Nosalewicz and Lipiec 2013). As root growth is expected to expand due to diminished soil moisture in the period of drought, the growth progression will also eventually show inhibition in root development or root expansion as earlier supported by Boyer (1970). Root growth inhibition in this case is also directly influenced by poor rates of or a lack of photosynthesis. Furthermore, low photosynthetic rates are also caused by the accumulation of abscisic acid (ABA) used to regulate stomatal conductance. The reduction of ABA keeps stomata open, while increased ABA concentration in the cells promotes stomatal closure. In turn, closed stomata limit gaseous exchange and lower the rate of photosynthesis.

Typically, water deficit causes alterations in plant metabolism affecting all cellular pathways, including cell division, elongation and expansion. The ions in the cells also become more concentrated leading to osmotic stress and the generation of ROS. Excess ROS cause damage to nucleic acids (RNA and DNA), disrupt membrane lipids, oxidise photosynthetic pigments and inhibit protein/enzyme activity.

11.4 GENERAL DEFENCE

The rapidly changing climate conditions make abiotic stresses a serious threat to sustainable plant life and agriculture. The world is experiencing increasingly fluctuating weather patterns that threaten the sustainability of the food production system globally. Escalating the problem is that plants are very slow in acclimatising to the ever-changing conditions. Typically, plants use a variety of mechanisms to respond to abiotic stress, such as alterations in growth pattern, plant morphology and physiological mechanisms. But these require a signalling response before plant transduction and acclimation pathways can be activated to deal with abiotic stresses. A variety of stress-sensing mechanisms include the following:

1) Mechanical influence of stress on plant or cell structure, like changes in plasma membrane as a result of stress;
2) Inhibition of enzymatic activities or denaturation of proteins;
3) Accumulation of unwanted by-products like ROS due to uncoupling of enzymatic or electron transfer reactions;
4) Signalling by stress-sensing proteins like the channel proteins;
5) Epigenetic sensing modifying DNA or RNA structures.

These signals trigger a downstream response comprising multiple signal transduction pathways involving a range of metabolites such as ROS, enzyme proteins and plant hormones, etc. It is this network of signals that will subsequently allow growth to continue during stressful conditions or turn off some pathways until more favourable conditions return. This process involves transcriptional regulators or transcriptional factors which are proteins that bind to specific DNA sequences and either activate or suppress the expression of other different genes. One such regulon occurring in soybean is the small ubiquitin-related modifier (SUMO) that is transcriptionally up-regulated by salinity stress, heat stress and ABA stimuli after at least 24 hours of plant exposure to stress. SUMO serves to regulate nuclear processes such as controlling transcription, subcellular trafficking and the cell cycle to homeostasis and biological activities during rapid environmental changes (Li et al. 2017).

Apart from signalling cascade and altered gene expressions that can be reversed when more favourable conditions arise, further genetic studies provided evidence showing that epigenetic changes may also provide long-term adaptation to abiotic stress. DNA methylation and histone modification have now been linked with specific abiotic stresses. Epigenetic factors cause phenotypic plasticity that enables plants to respond better to changing environments. Epigenetic variations have now proved to be trans-generationally heritable. Soybean epi-lines containing the MutSHOMOLOG-1 (MSH-1) system for inducing agronomically valuable epigenetic variation were reported. MSH-1 is a nuclear-encoded protein that influences mitochondrial and plastid properties, conferring tolerance to excessive light intensity, heat and drought stress (Raju et al. 2018).

Other studies revealed the use of endogenous hormonal regulations by plants to cope with abiotic stresses such as drought and flooding. Comparative RNA sequencing and hormonal profiling in soybean showed that ABA and ethylene responses were activated under both stresses. As transcriptomic-response analysis revealed, these hormones play an important role in coordinating transcriptomic energy-saving processes during abiotic stress (Tamang et al. 2021). The above study, as well as many others, demonstrated with conclusive evidence that plants also use hormones to mediate a wide range of adaptive responses and these hormones are essential in allowing plants the ability to rapidly adapt to abiotic stresses. Other ways used by legumes such as soybean to protect and defend themselves against abiotic stresses include osmotic adjustments by accumulating solutes and using them to lower water potential during the period of osmotic stress. In addition, soybeans may use antioxidants and ROS-scavenging pathways to protect plant cells from oxidative stress and the alteration of membrane lipids from flexible-liquid crystal structures to solid gel-like structures. Furthermore, they may use chaperons to protect essential proteins as well as the membrane structure from being damaged by abiotic stress constraints (Taiz et al. 2015; Mangena 2018).

11.5 RESISTANCE GENES FOR ABIOTIC STRESS TOLERANCE

Abiotic stress frequency varies over time, often occurring in combination or consecutively to particular stress conditions within the plant's environment. This causes a plant to have a multifaceted approach in responding to stresses. This plant stress-response is controlled by a complex network of signalling and regulatory pathways as indicated above, instead of using a 'one gene-one stress' approach. All plants including soybeans can face many abiotic threats, combined with biotic stress factors at the same time. This simply means that these plants need to contain genes making them tolerant to numerous stressful conditions. As Wang (2020) pointed out with regard to biotic stress, plants together with stress-inducing organisms need to co-evolve so that they can possess multiple resistance genes to deal with the rising virulence of pathogens, and enhanced tolerance to adverse conditions caused by environmental cues. It is also the responsibility of a plant breeder to constantly look for resistance genes.

To date, many genes and their associated molecular markers have been identified to deal with the ever-changing growth conditions. Similarly, research focus continues to test and investigate new plant mechanisms, as well as techniques that may be used to introduce such resistance genes into host plants, hoping to develop stress-resistant crop cultivars. QTLs for resistance against high-intensity ultraviolet-B (UV-B) irradiation in soybean were reported. Four QTLs, namely *UVBR12-1*, *UVBR6-1*, *UVBR10-1* and *UVBR14-1* resistance genes, were identified. The genomic regions containing *UVBR12-1* and *UVBR6-1* contained synthetic blocks that included other known resistance genes for both biotic and abiotic stress-related QTLs (Yoon et al. 2019). The UV-B serves as one of the most complex abiotic stresses that emanates from excessive plant exposure to high light intensity and heat stress, coupled with cell dehydration thereby negatively impacting on this crop's yield.

Furthermore, the transcriptomic analysis of calmodulin binding transcription activator (*GmCAMTA*) in soybean revealed multiple stress-related *cis*-regulatory elements (ABRE, SARE, G-box and W-box), 10 unique microRNA targets and 48 protein interaction networks. The *GmCAMTA* coding sequence was cloned into *A. tumefaciens* strain EHA105 recombinant vector (K599) and was expressed in soybean. This resulted in the promotion of drought-resistant hairy root development in overexpression lines in which the regenerated plants exhibited altered phenotypic, physiological and molecular characteristics (Noman et al. 2019). It is, therefore, very critical that QTL analysis and mapping continue in identifying stress-resistant traits and their associated genomic sequences. Understanding the composition of candidate abiotic or biotic stress-resistance genes in soybean is important for managing the diversity and divergence of such a gene pool which may have serious implications for plant breeding programmes in the future.

11.6 ABIOTIC STRESS-TOLERANT VARIETIES

Molecular studies revealed that signalling and regulatory proteins play a significant role in enhancing the growth and survival of plants living under stressful conditions. Many different transcription factor gene families (FTs) are being reported, serving a role in controlling gene expression, phenotype-related or physiological responses. Therefore, any observed variations in TFs may account for the different gene expression regulations at different stages of plant development, particularly in different tissues and organs (Wang 2020). Another family of regulatory proteins that are used for the protective functions in plants are cysteine proteases and their inhibitors. Cysteine protease inhibitors (also known as cystatins) exhibit inhibitory effects against endogenously produced phytocysteine proteases expressed due to biotic and abiotic stress (Misaka et al. 1996).

These proteolytic enzymes are abundant in all plants, including serine, aspartic and metallo-proteases that function as endopeptidases that cleave proteins distant from their termini into smaller fragments by aminolysing thioester bonds or carbonyl carbon of the reactive substrates. When soybean plants are exposed to abiotic stress conditions, proteolytic enzymes will be expressed to dismantle misfolding and damaged proteins, as well as to maintain turnover of cellular proteins as part of the overall attempt to protect the plant against stress. During this period of induced tissue senescence, proteins produced as a result of stress are degraded by proteases in order to remobilise nutrients or amino acids constituents either for repairing affected cells or from senescing tissues to other young developing tissues and organs (Mangena 2020).

Proteolytic enzymes cleave an unlimited number of peptide bonds generating free amino acids that subsequently produce fragments involved in secondary metabolite biosynthesis, and other harmful unwanted by-products that could generate ROS. These outcomes then led to the use of protease inhibitors (PIs) to regulate proteolytic enzyme activity of essential protein molecules during abiotic stress as an interesting strategy for the development of stress-resistant plants. Oryzacystatin-1 (*Oc-1*) and Oryzacystatin-2 (*Oc-2*) genes originating from rice (*Oryza sativa* L.) have been widely used for this purpose. Both *Oc*-genes successfully inhibited the activity of cysteine proteases induced due to biotic and abiotic stress. Such observations were also made in corn, wheat and many other cereal crops (Arai et al. 2002; Mahajan and Badgujar 2010).

So far, genetically engineered soybean varieties have also been developed through indirect *Agrobacterium* gene transfer harbouring a binary vector construct system containing the *Oc-1* gene used for tolerance against chilling and drought stress

(Mangena 2020). Other transgenic plants reported to express PIs included genetically transformed *Brassica*, *Nicotiana*, *Arabidopsis* and *Vigna unguiculata* against chilling, drought and pests (Sharma et al. 2000; Van der Vyver et al. 2003; Zhang et al. 2008; Prins et al. 2008). In other studies, transgenic soybean plants overexpressing the *Arabidopsis Δ1-pyrroline-5-carboxylate synthase* (P5CR) gene demonstrated elevated resistance to heat stress and drought by producing higher levels of proline (Homrich et al. 2012). Resilience of transgenic plants against abiotic stress by the expression of specific gene sequences was also reported in *Arabidopsis* ecotype Wassilewskija which exhibited increased tolerance to salt and osmotic stresses (Mazel et al. 2004).

11.7 SUMMARY

The information discussed in this chapter indicated that abiotic stress is responsible for the activation of many signalling and regulatory gene expressions. Clearly, plants employ these mechanisms with the intention of counteracting the stress and promoting their own survival. However, stresses like drought, chilling and salinity are responsible for significant yield losses in soybean on a worldwide scale. But naturally, it is these stresses which should drive the evolution, distribution and ecological fitness of plant species, including soybeans whose gene pool is very narrow. Overall, soybeans or most species within its family show a very high degree of vulnerability to different stressful environments. Only species such as *Vigna unguiculata* (cowpea) have the highest degree of adaptation to abiotic stresses through their modified morphological, physiological and reproductive traits. Delayed leaf senescence, stem greenness and deep rooting have been identified as traits that enhance cowpea's resistance to drought stress (Goufo et al. 2017).

Furthermore, the expression of metabolites such as proline, galactinol and quercetin correlate more with drought stress resistance in the roots of many legumes, including cowpea and soybean. As these plants acclimate to transient stressful environmental conditions by engaging multiple protective mechanisms, resistance genes must then be identified, characterised and incorporated into crop plants to achieve long-term adaptation. Drought appears to be the most limiting abiotic stress to soybean growth and productivity. Thus, recombinant proteins such as cysteine protease inhibitors originating from an *Oc-1* gene exhibiting inhibitory effects against proteolytic enzymes induced by drought or other type of stresses must be used to develop stress-resistant soybean plants. Phytocystatins have been long identified in *Oryza sativa* (rice), and have also been widely used for crop improvement or industrial applications. Drought is the most deleterious abiotic stress, and protective mechanisms involving stress escape and stress avoidance may not completely guarantee increase crop yield, except for the concept of drought stress tolerance.

Plants may briefly escape drought by changing their phenological development, activating early maturity or avoiding the stress by increasing root density or depth, closing stomata or rolling the leaves (Taiz et al. 2015). Although these protective measures may help maintain relatively high tissue moisture content and tissue hydrostatic pressure, they still have proved to be insufficient for the frequent levels of adverse abiotic stress. Scientific evidence shows that many regions across the globe have been experiencing transient severe drought events, flooding and warming temperatures as a result of climate change. These events cause too much pressure on crops' fitness and survival, and eventually this will continue being the case in the entire food production system.

11.8 ABBREVIATIONS

ABA	Abscisic acid
CL^-	Chloride ion
CO_2	Carbon dioxide
GmCAMTA	Calmodulin binding transcription activator
LEA	Late embryogenesis abundant protein
MSH-1	MutSHOMOLOG-1
Na^+	Sodium ion
Oc-1	Oryzacystatin-1
PI	Protease inhibitor
QTL	Quantitative trait loci
RNA	Ribonucleic acid
ROS	Reactive oxygen species
RR	Roundup ready
SUMO	Small ubiquitin-related modifier
TF	Transcription factor
UV-B	Ultraviolet-B

REFERENCES

Akpinar BA, Avsar B, Lucas AS and Budak H. (2012). Plant abiotic stress signalling. *Plant Signalling and Behavior* 7(11), 1450–1455.

Arai S, Matsumoto I, Emori Y and Abe K. (2002). Plant seed cystatins and their target enzymes of endogenous and exogenous origin. *Journal of Agricultural and Food Chemistry* 50(22), 6612–6617.

Battaglia M and Covarrubias AA. (2013). Late embryogenesis abundant (LEA) proteins in legumes. *Frontiers in Plant Science* 4(190), 1–11.

Boyer JS. (1970). Leaf enlargement and metabolic rates in corn, soybean and sunflower at various leaf water potentials. *Plant Physiology* 46, 233–235.

Bustingorri C and Lavado RS. (2011). Soybean growth under stable versus peak salinity. *Scientia Agricola* 68(1), 102–108.

Goufo P, Moutinho-Pereira JM, Jorge TF, Correia CM, Oliveira MR, Rosa EAS, Antonio C and Trindade H. (2017). Cowpea (*Vigna unguiculata* L. Walp.) Metabolomics: Osmoprotection as a physiological strategy for drought stress resistance and improved yield. *Frontiers in Plant Science* 8(586), 1–22.

He M, He C-Q and Ding N-Z. (2018). Abiotic stresses: General defenses of land plants and chances for engineering multistress tolerance. *Frontiers in Plant Science* 9(1771), 1–18.

Homrich MS, Wiebke-Strohm B, Weber RLM and Bodanese-Zanettini MH. (2012). Soybean genetic transformation: A valuable tool for the functional study of genes and the production of agronomically important plants. *Genetics and Molecular Biology* 35(4) Supplement 1, 998–1010.

Li Y, Wang G, Xu Z, Li J, Sum M, Guo J and Ji W. (2017). Organisation and regulation of soybean SUMOylation system under abiotic stress conditions. *Frontier in Plant Science* 8(1458), 1–14.

Los DA and Murata N. (2004). Membrane fluidity and its role in the perception of environmental signals. *Biochimica et Biophysica Acta (BBA)- Biomembranes* 1666(1–2), 142–156.

Mahajan RT and Badgujar SB. (2010). Biological aspects of proteolytic enzymes: A review. *Journal of Pharmacy Research* 3(9), 2048–2068.

Mangena P. (2018). Water stress: Morphological and anatomical changes in soybean (*Glycine max* L.) plants. In Andjelkovic V (eds), *Plant, Abiotic Stress and Responses to Climate Change*. IntechOpen, London. pp. 9–31.

Mangena P. (2020). Phytocystatins and their potential application in the development of drought tolerance plants in soybean (*Glycine max* L.). *Protein and Peptide Letters* 27, 135–144.

Mansour MMF. (2013). Plasma membrane permeability as an indicator of salt tolerance in plants. *Biologia Plantarum* 57(1), 1–10.

Mazel A, Leshem Y, Tiwari BA and Levine A. (2004). Induction of salt and osmotic stress tolerance by overexpression of an intracellular vesicle trafficking protein AtRab7 (AtRabG3e). *Plant Physiology* 134, 118–128.

Misaka T, Kuroda M, Iwabuchi K, Abe K and Arai S. (1996). Soyacystatin, a novel cysteine proteinase inhibitor in soybean, is distinct in protein structure and gene organization from other cystatins of animal and plant origin. *European Journal of Biochemistry* 240(3), 609–614.

Noman M, Jameel A, Quang W-D, Ahmad N, Liu W-C, Wang F-W and Li H-Y. (2019). Overexpresion of GmCAMTA12 enhanced drought tolerance in *Arabidopsis* and soybean. *International Journal of Molecular Sciences* 20(4849), 1–24.

Nosalewicz A and Lipiec J. (2013). The effect of compacted soil layer on vertical root distribution and water uptake by wheat. *Plant and Soil* 375, 229–240.

Pegg DE. (1987). Ice crystals in tissues and organs. In Pegg DE and Karow AM (eds), *The Biophysics of Organ Cryopreservation, NATO ASI Series (Series A: Life Sciences)*, Vol 147. Springer, Boston. pp. 117–140.

Prins A, Van Heerden PDR, Olmos E, Kunnert KJ and Foyer CH. (2008). Cysteine proteinases regulate chloroplast protein content and composition in tobacco leaves: A model for dynamic interactions with ribulose-1,5-bisphosphate carboxylase/oxygenase (Rubisco) vesicular bodies. *Journal of Experimental Biology* 59(7), 1935–1950.

Raftery AE, Zimmer A, Fierson DMW, Startz R and Liu P. (2017). Less than 2°C warming by 2100 unlikely. *Nature Climate Change* 7, 637–641.

Raju SKK, Shao M-R, Sanchez R, Xu Y-Z, Sandhu A, Graef G and Mackenzie S. (2018). An epigenetic breeding system in soybean for increased yield and stability. *Plant Biotechnology Journal* 16(11), 1836–1847.

Sharma HC, Sharma KK, Seetherama N and Ortiz R. (2000). Prospects of using transgenic resistance to insects in crop improvement. *Electronic Journal of Biotechnology* 3(2), 76–95.

Taiz L, Zeiger E, Moller IM and Murphy A. (2015). *Plant Physiology and Development*, 5th ed. Sinauer Associates, Sunderland, Massachusetts. pp. 731–761.

Tamang BG, Li S, Rajasundaram D, Lamichhane S and Fukao T. (2021). Overlapping and stress-specific transcriptomic and hormonal response to flooding and drought in soybean. *Plant Journal* 107(1), 100–117.

Van der Vyver C, Schneidereit J, Discoll S, Turner J, Kunnert K and Foyer CH. (2003). Oryzacystatin 1 expression in transformed tobacco produces a conditional growth phenotype and enhances chilling tolerance. *Plant Biotechnology Journal* 1(2), 101–112.

Wang RRC. (2020). Chromosomal distribution of gene conferring tolerance to abiotic stresses versus that of genes controlling resistance to biotic stresses in plants. *International Journal of Molecular Sciences* 21(1820), 1–9.

Waqas MA, Kaya C, Riaz A, Farooq M, Nawaz I, Wikes A and Li Y. (2019). Potential mechanism of abiotic stress tolerance in crop plants induced by thiourea. *Frontier in Plant Science* 10(1336), 1–14.

Wesley-Smith J, Berjak P, Pammenter NW and Walters C. (2014). Intracellular ice and cell survival in cryo-exposed embryonic axes of recalcitrant seeds of *Acer saccharinum*: An ultrastructural study of factors affecting cell and ice structures. *Annals of Botany* 113(4), 695–709.

Yoon MY, Kim MY, Ha J, Lee T, Kin KD and Lee S-H. (2019). QTL analysis of resistance to high-intensity UV-B irradiation in soybean (*Glycine max* [L.] Merr.). *International Journal of Molecular Sciences* 20(3287), 1–15.

Zhang H, Zhu J, Gong Z and Zhu J-K. (2021). Abiotic stress responses in plants. *Nature Review Genetics*, *23,* 1–16.

Zhang X, Liu S and Takano T. (2008). Two cysteine proteinase inhibitors from *Arabidopsis thaliana*, AtCYSa and AtCYSb, increasing the salt, drought, oxidation and cold tolerance. *Plant Molecular Biology* 68(1–2), 131–143.

12 Potential Health Safety Concerns and Environmental Risks

12.1 INTRODUCTION

Advancements made more than three decades ago have led to the use of microorganisms and plants to revolutionise plant molecular genetics and breeding. Biotechnology's most successful applications such as the development of transgenic crops (Table 12.1 and 12.2), recombinant proteins, CRISPR-based gene editing, transgenic animals, microarrays, adoptive cell therapy and next-generation DNA sequencing are among some of the most critical approaches in human civilisation. These progressive developments have undoubtedly given birth to a whole new industrial technology that is dedicated to the development, manufacture, use and marketing of biotechnological products, especially those that are aimed at improving and meeting the greater needs of the human population.

However, some applications such as genetic engineering of plants (also known as plant transformation) using bacterial vectors, like *Agrobacterium tumefaciens*, have played a key role in the production of genetically modified organisms (GMOs) (Table 12.1 and 12.2). These products include recombinant proteins, the recovery of fertile transgenic plants and the synthesis of useful secondary metabolites. Plant transformation forms a major part of a very diverse green biotechnology that continues to provide ample opportunities for establishing new measures aimed at combating food insecurity, poverty and climate change. Genetic transformation of both monocot and dicot plants (Table 12.1 and 12.2) using *A. tumefaciens* is reported as the most rapid and efficient foreign DNA transfer system. But this technology is also accompanied by many challenges involving protocol inefficiencies, lack of proper genotype specificity controls and the public scepticism/criticism surrounding the commercialisation and consumption of these bioengineered products.

Agrobacterium-mediated genetic transformation accounts for the largest number of market-available transgenic soybean and other crop varieties. The newly improved varieties were developed using both modified and standard transformation procedures, which usually involve the transfer and integration of the T-DNA contained within a binary plasmid vector construct. As previously discussed, the binary vectors used also carry a *bar* gene driven by the *uidA* promoter gene and the virulence (*vir*) genes. As discussed in Chapter 1, virulence genes (*virA*, *virB*, *VirC vir D*, *virE* and *VirG*) are used by the *Agrobacterium* to sense and recognise signals emitted by the host plant tissues to enable the introgression of genes of interest into

DOI: 10.1201/b22829-12

TABLE 12.1
List of monocotyledonous plants successfully transformed using *Agrobacterium tumefaciens*-mediated genetic transformation

Species	Culture/explant	*Agrobacterium* strain	Reference
Allium sativum L	Callus	EHA101	Koetle et al. (2015)
Agapanthus praecox Willd	Callus	EHA101	Koetle et al. (2015)
Lolium perenne L.	Callus	EHA105	Zhang et al. (2013)
Gladiolus primulinus	Shoot tips	LBA4404	Koetle et al. (2015)
Hordeum vulgare L.	Immature embryos	LBA4404/ pSBI	Marthe et al. (2015)
Lilium sp. cv Acapulco	Callus	EHA101	Koetle et al. (2015)
Oryza sativa L.	Callus/ embryos	EHA101	Hiei et al. (1994)
Hyacinthus orientalis L.	Leaf pieces	CBE21	Koetle et al. (2015)
Sorghum bicolor L.	Immature embryos	AGL1	Wu et al. (2014)
Avena sativa L.	Immature embryos/ leaf explants	LBA4404	Gasparis et al. (2008)
Lilium longiflorum Thunb.	Bulb scales	EHA101	Koetle et al. (2015)
Narcissus tazetta L.	Leaf pieces	LBA4404	Koetle et al. (2015)
Ornithogalum thysoides Jacq.x	Leaf pieces	AGL1	Koetle et al. (2015)
Tricyrtis hirta Thunb	Callus	EHA101	Koetle et al. (2015)
Tricyrtis wali. sp. shinonome	Callus	EHA101	Koetle et al. (2015)
Typha latifolia L.	Callus	EHA101/ LBA4404	Koetle et al. (2015)
Zea mays L.	Immature embryos	LBA4404	Ishida et al. (2015)
Pennisetum glaucum L.	Shoot apex callus	LBA4404	Ignacimuth and Kannan (2013)
Saccharum officinarum L.	Axillary buds	EHA105	Manickavasagam et al. (2004)

hosts. Therefore, a small number of *Agrobacterium* strains have been used for the transformation of monocotyledonous and dicotyledonous plants as indicated in the tables (Table 12.1 and 12.2).

The strains containing super-virulent tumour-inducing plasmids possessing wider host range and high transformation efficiency that include AGL0, AGL1, A281, LBA4404, EHA101 and EHA105 have been widely used (Hiei et al. 2014). These strains are popularly known for their efficacious transformation frequencies in cereal crops as compared to dicot plants like soybean, as demonstrated in Table 12.1 and 12.2. Although monocots were believed to be not transformable, it was eventually illustrated that these species are more amenable to transformation than most dicots, especially when compared to recalcitrant soybean genotypes. The most highly efficient transformation rates were reported in rice (*Oryza sativa* L.) and maize (*Zea mays* L.) using *Agrobacterium* co-cultivated immature embryos in callus cultures (Hiei et al. 1994; Ishida et al. 1996). Currently, monocot transformation protocols are well established, and agronomically important species such as barley (*Hordeum vulgare* L.), sugarcane (*Saccharum officinarum* L.), sorghum (*Sorghum bicolor* L.) and wheat (*Triticum aestivum* L.) are easily transformable using *A. tumefaciens.*

TABLE 12.2
List of dicotyledonous plants successfully transformed using *Agrobacterium tumefaciens*-mediated genetic transformation

Species	Culture/explant	*Agrobacterium* strain	Reference
Gossypium hirsutum L	Callus	LBA4404	Wu et al. (2008)
Cajanus cajan L.	Embryogenic axis-attached cotyledons	EHA101	Karmakar et al. (2019)
Phaseolus vulgaris L.	Embryogenic axis	LBA4404	Solis-Ramos et al. (2019)
Vicia faba L.	Internodal segments	GV3850	Opabode (2006)
Brassica napus L.	Hypocotyl/ mesophyll protoplast	GV850	Opabode (2006)
Cicer arientum L.	Embryo axes	C58C2/GV2260	Krishnamurthy et al. (2000)
Glycine max L. Merrill	Cotyledonary node	C58/AGL1/EHA101/EHA105/LBA4404	Olhoft et al. (2001)
Arabidopsis thaliana	Petiole explants	C58C1Rif	Chateau et al. (2000)
Arachis hypogea	Leaf explants	LBA4404	Eapen and George (1994)
Brassica oleracea var. botrytis	Cotyledons	LBA4404	Bhalla and Smith (1998)
Eucalyptus camaldulensis	Hypocotyls	EHA101	Ho et al. (1998)
Malus domestica	Leaf pieces	EHA101	De Bondt et al. (1996)
Pinus sp.	Zygotic embryos	LBA4404	Tang (2001)
Cleome gynandra L.	Hypocotyls/ cotyledons	LBA4404	Newell et al. (2010)
Brassica oleracea var. italica	Peduncle/ seedlings	ABI	Metz et al. (1995)
Brassica oleracea var. capitate	Seedling explants	ABI	Metz et al. (1995)
Camellia sinensis (L.) O. Kuntze	Embryogenic tissues	LBA4404	Jeyaramraja and Meenakshi (2005)
Solanum melongena L.	Root	LBA4404	Franklin and Sita (2003)

However, researchers continue to worry about the DNA transfer complexity and specificity of susceptible plant genotypes, conditions of explants, cultures, selective agents, type of strains used and other technical hurdles affecting the transformation process in dicot species. Meanwhile, less emphasis is predominantly placed on the influence of this technology on the environment. More serious issues involving the concerns raised by the public regarding possible environmental risks that are associated with the use of transgenic crops have also gained momentum. Therefore, the purpose of this chapter is to scrutinise these issues, including other potential effects on the health and wellbeing of humans and animals.

12.2 EVIDENCE OF BIOSAFETY AND RISKS INVOLVING *AGROBACTERIUM TUMEFACIENS*

The DNA fragments transferred via *Agrobacterium* vectors carry a number of genes that are transcribed and translated in plant cells. Genetic studies have demonstrated that the products of these genes are responsible for the abnormal growth patterns of 'crown gall' and 'hairy root' seen in plants. As indicated in Chapter 2, Ti-plasmid is the causative agent for the 'crown gall' disease; meanwhile Ri-plasmid is responsible for the 'hairy root'. The characterisation of T-DNA linked genes revealed that oncogenes' products are directly involved in causing the abnormal plant cell growths. Furthermore, as Gelvin (2003) stated, the proliferation of undifferentiated cells in 'crown gall' tumours arises from the production of two major plant hormones, namely auxins and cytokinins. These plant hormones are shown to occur as a consequence of T-DNA gene transcriptions and translations.

T-DNA gene 1 (*iaaM*) and gene 2 (*iaaH*) function to code for tryptophan 2-monooxygenase that catalyses the conversion of tryptophan to indole-3-acetamide, and an indole-3-acetamide hydrolase which catalyses the formation of indole-3-acetic acid from indole-3-acetamide, respectively (Koncz et al. 1992). In addition, the abnormal production of cytokinins by gene 4 (*iptZ*) coding for isopentenyl transferase catalysing adenosine-5-monophosphate (5'AMP) and isopentenyl phosphate into an active cytokinin isopentenyladenosine-5-monophosphate. However, not only can the hormone-induced abnormal phenotypes that were observed be correlated with T-DNA gene expressions, but some evidence emerged of biosafety and risks associated with these *Agrobacterium* genes. A detailed analysis regarding this issue remains very critical as transgenic plants directly or their products are used to argument conventional breeding by improving agricultural practices, thereby achieving enhanced yields and quality of crops.

Conversely, the potential horizontal gene transfers to non-targeted hosts, environmental genetic pollution and undesirable gene flows to closely related species remain major causes of criticisms against plant genetic engineering. Furthermore, this tool faces more denigrations because of its advancements that have made it possible to transfer and express genes into crops that are from unrelated plants and even from non-plants organisms. Such unintended consequences of genetic engineering are discussed below.

12.3 UNINTENDED T-DNA AND SELECTABLE MARKERS TRANSFER

The incorporation of some T-DNA and selectable marker genes into genetically modified products is highly undesirable for consumers. Advancements in plant transformation have made it possible for genetic engineers to use bacteria and viruses as genetic reservoirs for genes of interests (*goi*). Although researchers thoroughly investigated and elucidated the functions of these genes, transgenic products that contain and express genes obtained from sources other than plants are still heavily criticised. Historically, the production of transgenic plants using improved agronomic traits depended on the use of genes derived from plants containing stress-resistance genes.

It was reported that plants containing modified acetolactate synthase (ALS) gene derived from Poaceae (*Zea mays* and *Hordeum vulgare*) and Brassicaceae (*Brassica napus*) displayed the same high levels of sulfonylurea tolerance as transgenic plants expressing the bacterial ALS tolerance genes (Barampuram and Zhang 2011). These observations have led to the identification of a large number of stress-resistant genes that are presents in plants, whereby many of them are still identified and characterised through QTL and other functional genomics-based programmes. Several of these identified genes are currently being used as viable alternatives to the use of microbial genes for crop improvement. Such genes include glyphosate tolerance gene and disease-resistant genes as indicated in Table 12.3. The concerns raised, however, elaborate that genetic transformation using genes native to plants without selectable markers will be highly desirable to consumers rather than the opposite.

Researchers should consider and further investigate the use of marker genes isolated from plants. However, these and marker-free techniques available for use still show limitations and the gene system is not yet well-established (Rosellini 2011).

TABLE 12.3
Examples of functionally active resistant genes derived from plant sources instead of any other microbial, viral or animal source

Plant genes	Source/species	Reference
Hm1	*Zea mays* L. (maize)	Takken and Joosten (2000)
T-urf13	*Zea mays* L. (maize)	Chaumont et al. (1995)
Mlo	*Hordeum vulgare* (barley)	Takken and Joosten (2000)
oc-1	*Oryza sativa* L.	Mangena et al. (2015)
Pto, *Cf-2* and *Cf-9*	*Lycopersicon esculentum*	Jones (1996)
N	*Nicotiana tabacum*	Jones (1996)
L6	*Linum* spp.	Jones (1996)
RPS2 and *RPM1*	*Arabidopsis thaliana*	Jones (1996)
Xa21	*Oryza sativa* L.	Jones (1996)
Bs1, *Bs2* and *Bs3*	*Capsicum annum*	White et al. (2009)
CYR1	*Vigna mungo*	Zvereva and Pooggin (2012)
Rx1, *Rx2*, *Y-1* and *Tm-2*	*Solanum tuberosum*	Zvereva and Pooggin (2012)

Most of these plant resistance genes have not been thoroughly investigated, characterised and their phenotypes examined. The efficiency of bacterial resistant genes enabling the selection of transformed plant cells from non-transgenic ones or even in foods is generally believed to pose no safety threats by consumers. However, the scepticism surrounding the use of these genes by the general public and regulatory agencies is still problematic, even though there is no concrete evidence attributing these selectable marker genes or any T-DNA genes to any current or emerging health or environmental problems.

12.4 GENE TRANSFER TO NON-TARGETED HOSTS

Horizontal gene transfer (HGT) is one of the processes involved in acquiring new gene combinations in both unicellular and multicellular organisms. The process serves as a major driving force leading to genomic variability contributing to adaptation and speciation where offspring could contain the newly transferred genetic materials. The integration of genetic materials between different strains or species is often non-transient and very stable. This transfer can affect niche speciation, resistance to diseases, cause metabolic shifts and generate 'hotspots of superbugs'. There are several reports indicating that this phenomenon is involved in the development of antibiotics, antifungal and antiviral drug resistance genes in microorganisms (Burmeister 2015; Raghavendra and Pullaiah 2018; Emamalipour et al. 2020).

Existing evidence suggest that the genes responsible for this kind of multidrug resistance, especially as observed in bacterial resistance to antibiotics, are acquired in the course of HGT. The horizontal acquisition of antibiotic resistance and virulence genes has already caused serious disease outbreak leading to major morbidity and mortality in humans and animals. Various researchers including Burmeister (2015) demonstrated these resistance effects in relation to human medicine and agriculture. The report analysed the selective environmental response of bacteria that occurred intetracycline- and methicillin-fed animals which caused resistance to antibiotics that was transferred to the novel human-associated *Staphylococcus aureus* strain CC398. Other investigations also reported the presence of homologous sequences of the Ri-plasmid of *Agrobacterium rhizogenes* (cT-DNA) in the genome of untransformed *Nicotiana glauca*.

Although this appeared over 30 years ago, the expression of both oncogene and opine synthesis genes indicated that they may still play a critical role in the evolution of *Nicotiana* spp. Similar observations were later reported in *Linaria vulgaris* and *Linaria dalmatica*, indicating the creation of new plant species from T-DNA oncogene and the *MIS* gene insertions (Matveeva and Lutova 2014). Undoubtedly *Agrobacterium* is considered a natural biological engineer because it is well equipped with a full set of virulence genes and entire enzyme machinery required for both induced and horizontal gene transfer. But, if this process is left unmonitored, the wide host range found in this bacterium can mediate unintended plant genetic transformation as well as the transfer of genes in other closely related and unrelated organisms.

12.5 UNDESIRABLE GENE FLOW TO CLOSELY RELATED SPECIES

An in-depth examination of the potential impact and importance of gene flow from *Agrobacterium*-transformed species to other untransformed organisms is also an important topic. Studies involve non-GMO species closely related to the transformed soybean plants, in addition to organisms like fish and animals. One of the main concerns in this regard is the unanticipated cross pollination/gene transfer that may take place between genetically modified and untransformed compatible species. This is said to be one of the main issues as far as gene pollution by transgenic crops is concerned. However, not much is currently known about the degree or levels of cross fertilisation between species, or among some of the widely cultivated genetically engineered crops such as soybean, rice, maize and their non-GMO counterparts.

Galeano et al. (2010) reported the assessment of the occurrence and frequency of cross-pollination between commercial transgenic and non-transgenic maize crop varieties in Uruguay. They reported adventitious presence of transgenic materials in non-GM crop offspring. The foreign genetic materials were found in at least three of the five cases reported. This analysis identified events of putative sources of transgenic pollen that was found in non-transgenic seedling maize plants, with estimated frequencies of 0.53, 0.83 and 0.13% in the three identified varieties. This is one of the few reports that clearly indicated that the advent of GM crops cannot be reversed, and their potential risks that are directly associated with transgenic pollen should be prevented or reduced immediately. The isolation of GM crops by distance or time may serve as one of the simplest and most effective methods used for the prevention of foreign gene transfers through HGT.

Viljoen and Chetty (2011) suggested the combination of both temporal and distance isolations, together with the consideration of transgenic pollen sources to make these preventative measures more applicable and practical. The undesirable gene flow among GM and non-GM maize has been widely reported, especially under commercial farming. Maize and soybean in particular, apart from grains/oilseed crops mentioned in Tables 12.1 and 12.2, remain the most controversial and extensively cultivated genetically modified crops. Furthermore, this controversy about the potential risks of unintended gene flow from GMOs such as maize and soybean to non-GM populations and the negative impacts they will impose on biodiversity still remain centred around both maize and soybean GMOs. This is so because of the fact that there is not much commercial-scale cultivation of GM trees or GM fish farming or animals that are being produced for human consumption.

Nevertheless, the debate has not improved or shifted as it remains controversial and polarised with recently emerging terminologies such as 'genetic pollution' being introduced, often without concrete scientific facts. These emergent terms also create confusion among consumers and law makers due to the lack of clarity or scope. It is well known that a directly correlated relationship has been in existence between naturally occurring gene flows and evolution, where gene flows in plants have been considered as a cohesive force of evolutionary significance. Therefore, more research on the interactions between GM and non-GM populations is required in order to improve the understanding of the topic. The knowledge generated will be useful for characterising the role of gene flows in evolution for species fitness, and to create a neutral, moderated and vigorous review platform for this debate.

12.6 POTENTIAL RISKS TO HUMAN AND ANIMAL HEALTH

A million-dollar question is whether transgenic materials generated through techniques such as *Agrobacterium*-mediated genetic transformation will, if not yet, in future, affect human and animal health. As a precautionary measure, genetic engineers always conduct additional studies prior to GMO release and often set up buffer zones to prevent external gene flows (Rizwan et al. 2019). In crop plants, pollen-mediated gene flow was described as a major pathway of undesirable GM gene transfer to non-GM counterparts and wild relatives. But so far, there is no mention of gene flow into human and animals. However, GMO safety remains a bone of contention globally despite their immense benefits that have been put forward. These have always been considered a threat to human and animal health, bringing along greater publicity and public scrutiny. Furthermore, GMOs are seen as a threat to the human race despite the fact that they are efficiently used to eradicate poverty and hunger, and may be used to effectively counteract the effects of climate change such as drought, salinity and high temperatures.

It is, therefore, necessary for government and regulatory bodies to provide evidence of whether GMOs or their products could have harmful consequences for human or animal lives. Prakash et al. (2011) indicated that microorganisms that are genetically improved have the ability to reproduce and establish themselves as persistent populations, posing long-term effects on other biological communities. This often becomes the case because attained genetic manipulations may consequently change other characteristics of the transformed organisms apart from those of the replaced or inserted genes. However, it is still quite challenging for researchers, regulators and critics to tell whether these organisms or the consumption of GMO-derived foods will make people and animal sick or live a few years less than if they never consumed them. A major anticipated risk for consumers was that changes made in the genetically engineered crops may result in unwanted toxins or allergen production in foods such as cupin and gluten (Figure 12.1).

But GM foods available for public consumption have passed detailed risk assessments of possible allergenicity and toxicity prior to attaining marketing approvals (Lee et al. 2017). The assessments include an evaluation of the protein encoded by the transgene, as well as a risk assessment of the crop that incorporated this gene of interest. Such evaluations take into account the considerations of safety and history of exposure of the gene source, protein structure or amino acid sequence identified as allergens and protein stability to many clinical and biochemical evaluations (Ladics 2018). The basic crystal structures of cupin and prolamins superfamilies serving as the most widespread group of protein allergens are indicated in Figure 12.1.

These crystal structures of cupin protein (tm1459) in copper (Cu) substituted form, gluten prolamins and non-gluten prolamins are proclaimed as GMO allergens as reported by Fujieda et al. (2020), Balakireva and Zamyatnin (2016) and Espinoza-Herrera et al. (2021). The 7S globulins (beta-conglycinin) and glycinin are common

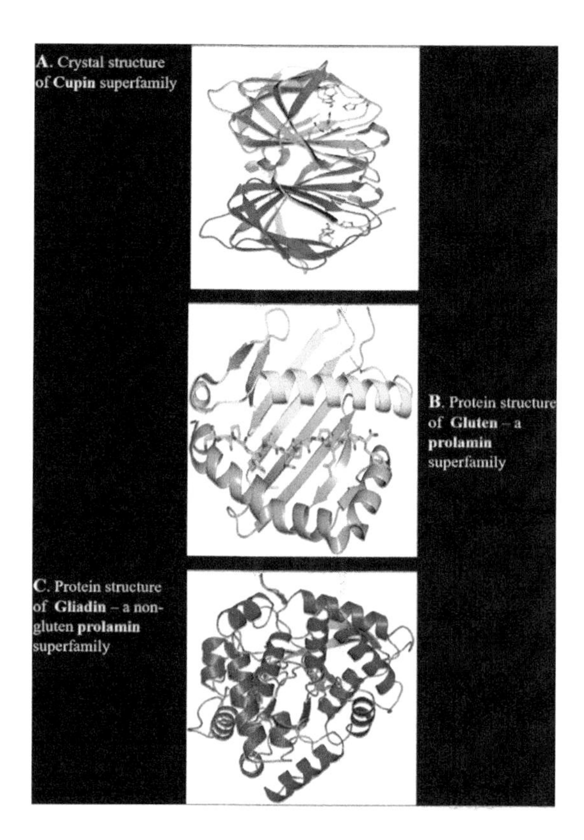

FIGURE 12.1 Allergenic protein structures of cupin and two prolamins (gluten and gliadin). The prolamin superfamily contain a conserved skeleton of eight cysteine residues within their protein sequences (Breiteneder and Radauer 2004).

allergenic seed storage cupins found in soybean. Unlike cupins, prolamins as the second most abundant storage proteins in grasses can be found in wheat (gliadin), barley (hordein), rye (secalin), corn (zein), sorghum (kafirin) and oats (avenin). A prevalence of these allergenic proteins is a serious problem as they are recognised by the immune system as foreign and dangerous substances causing allergic symptoms (Tsuji et al. 2001; Mills et al. 2002).

12.7 ENVIRONMENTAL RISK ASSESSMENT AND CONTAMINATION

The major environmental concerns associated with GM crops are said to be their possible negative effects on organic farming, seed production and conservation of natural genetic resources, among others. An increase in the amount of GM crops also has the potential to create new weeds through outcrossing with wild relatives and by simply becoming super weeds themselves. However, the adoption of insect- and herbicide-resistant GM crops has reduced excessive pesticide and insecticide spraying to those crops. The use of GM crops results in reduced tillage and use of heavy machinery that consumed high amounts of fuel during cultivation. This has also significantly reduced greenhouse gasses emissions especially in areas where GM cropping is expanded (Brookes and Barfoot 2015). As previously highlighted, the cultivation of transgenic crops containing changes in the overall pattern of gene expression is more feared due to its potential to escalate 'genetic pollution' between genetically transformed and non-GMOs.

Prakash et al. (2011) suggested that 'genetic pollution' may arise from gene sequence rearrangement or deletions that may cause some potential gene instability and interference with other gene functions possibly causing risks. Such risks to the open environment are summarised in Table 12.4. Currently, the evidence or insights of direct hazardous effects of GMOs on the environment as suggested in the table are conflicting. Ample available data provide scientifically assessed proof of lesser environmental risks emanating from the use of GMOs. Although the consequences of gene flows are debatable, no concrete evidence has attributed any environmental risks to the adoption and use of GMOs. According to Tsatsakis et al. (2017), adequate and precise scientific evidence provided to regulatory agencies, governments, policy makers and agribusiness is vital to enable them to thoroughly investigate any possible risks of using genetically transformed organisms/products.

TABLE 12.4
A list of ten potential environmental risks caused by the use of GMOs, derived by the insertion of a single/multiple genes from animal, plant or microbial sources

1. Genetic contamination by interbreeding/hybridisation with the wild type or other sexually compatible relatives.
2. Invasive nature, spreading into new habitats and causing ecological damage.
3. Altered seed purity resulting from genetic contamination or pollen drift from GMO fields.
4. Impairment of the natural gene pool or natural biodiversity.
5. Risk of the development of super competitive weeds that may destroy other forms of life.
6. The risk of evolution of common plant bacteria, fungi and viruses to become more resistant to the currently used antimicrobial agents or form new mutant strains.
7. The risk of evolution of common pests to becoming more resistant/less susceptible to pesticides.
8. Harm to non-targeted insects, birds and other animals.
9. Increased resistance to natural selection pressure, causing evolution of distinct resistant populations.
10. Disruption of biological agroecosystems or biotic communities and their associations with other organisms, e.g., parasitism or symbiosis.
11. Ecological damage and destruction at species/population level due to competition/fitness.
12. Undesirable horizontal gene transfer or lateral gene transfer between unicellular and multicellular organisms. This includes, for example, the transfer of antimicrobial resistance genes to pathogenic viruses or bacteria.

Sources: Prakash et al. (2011), Kramkowska et al. (2013) and Fontes et al. (2002).

12.8 MICROBIAL COMPETITION WITH TRANSFORMED *AGROBACTERIUM* STRAINS

The natural environment contains microorganisms that are found in specific major domains of Archaea, Eukarya and Bacteria with the Eukaryotic domain constituting microbial fungi, protozoa and algae. In addition to these microbes, viruses also form part of the other important biological entities infecting both prokaryotic and eukaryotic cells. Pepper and Gentry (2015) also alluded to the infectious single-stranded RNAs (viroids) that are found in the environment and infect plants. Both viroids and viruses rely upon the infection of their hosts to reproduce and sustain their life cycles, since they do not possess any metabolic capacity. The populations of all of these organisms inhibit the same environment where they co-exist or compete for nutrients and other resources. On the other hand, competition between two populations may be severe resulting in the elimination of one population by another, often through the excretion of chemicals that are toxic or inhibitory into the environment (Fredrickson and Stephanopoulos 1981).

Stringent risk assessment is required for engineered microorganisms that are released to the environment for use in bioremediation, mining, control of other harmful bacteria, viruses and to alleviate crop damage by environmental cues. For instance, a *Pseudomonas syringae* strain called the 'ice-minus' was released to the environment to putatively protect crops from frost damage by preventing the formation of ice crystal in the tissues (Luthy 1991). Currently, no dangers have been reported on the effects of engineered *Agrobacterium* following any potential releases to the natural environment. But there are expectations that potential threats may succeed and the potential hazards may occur following the continued use of transformed *A. tumefaciens* strains in the environment. The types of escapes that are anticipated may include unintended vertical gene flow (VGF) and HGF. VGF refers to the transfer of genes or alleles by normal reproductive means between different species. Meanwhile, as previously highlighted, HGF refers to the movement of genes by means other than reproductive processes.

Pollen-mediated gene flow, seed-mediated gene flow and vegetative-propagule-mediated gene flow control the unintended movements of genes between separate plant populations (Lu 2008). These observations were not surprising since risk assessment in the case of engineered microorganisms and their possible competition with other bacteria found in natural habitats is challenging. Furthermore, the said challenges are exacerbated by the fact that microbial effects on the environment are not easily observable. Microbial effects are difficult to predict given the fact that they possess exponential growth, and variably exchange genetic materials either through HGF or VGF, so rapidly. However, *A. tumefaciens* is not native to many environments and host plant species. With the considerable *Agrobacterium* diversity of strains that exist, including the different characteristics that similar strains living in different environments present, many of these features result in their competition and risk being assessed based on species or strain differences.

12.9 IMPACT OF *AGROBACTERIUM* ON THE ENVIRONMENT

Agrobacterium-mediated gene transfer now represents a sector in agricultural biotechnology. According to Luthy (1991), this technology combines natural sciences and engineering sciences in order to achieve the production of goods and services for the benefit of mankind. As already highlighted, gene technology may have unintended adverse effects on the environment, communities and ecosystems at large. The escape of engineered *Agrobacterium* strains may negatively affect different populations within the ecosystem due to the fact that all microorganisms are self-replicating. This means that it may be impossible to control their spread simply by discontinuing their further release to the environment. As suggested by Lenski (1993), effects of engineered *Agrobacterium* may be minimal due to the changes in their genetic makeup, disruption of genomic coadaptation and energetic inefficiencies. These strains are reported to be less fit than their progenitors and with reduced competitive fitness as a result of their genetic modifications.

The continuing challenges faced by researchers in effectively assessing the risks associated with genetically engineered microorganisms (GEMs) mean that the engineered genes will remain a complex ecological and evolutionary problem (Lenski 1993; Lu 2008). The question therefore remains of how researchers and environmental agencies could predict, manage and eradicate any potential negative effects of GEMs on the environment without compromising the positive role they have in sustainable agriculture, aquaculture, forestry, bioremediation and other essential ecological services. Analysis of the ecological consequences of *Agrobacterium*-environment interactions is important to counteract problems highlighted in Table 12.4, and to clarify whether these microbes indeed cause irreparable loss in species diversity or genetic diversity within species without any positive role that they play in this process.

12.10 SUMMARY

For more than two decades, scientists have used *Agrobacterium tumefaciens* gene transfer technology for the improvement of cereals and pulses for food, feed and other applications. Relatively few monocots and dicots species have been successfully genetically manipulated using this bacterium. However, it was discussed in this and other chapters that the technology involves complex microbial genetic manipulations for the insertion of the genes of interest (*goi*) into Ri- or Ti-plasmids. The introduction of *goi* derived from sources (viruses, bacteria, animals, etc.) other than plants for applications in food and feed crops implies that there are major gene additions, and perhaps specific concerns. For example, the development of antibiotic resistance in pathogenic microorganism, especially those derived from unintended gene flows between GM and non-GM crops, would potentially cause serious problems in human medicine.

Nap et al. (1992) attributed these new characteristics of bacterial resistance to the high antibiotic selection pressure resulting from the use of kanamycin resistance and the *eu-phA2* genes. These genes contribute to the unacceptable increase in resistance of potentially pathogenic microorganisms and impair the human as well as veterinary applications of these antibiotics. Despite a lot of research efforts, the presence and biological effects of genetically engineered *Agrobacterium* on the environment are not yet well researched. This excludes the role of genes transferred by their bacterial vectors to crop plants and the escape of such genes into the environment. Risk analysis indicates that any potential ecological role of transformed *Agrobacterium* strain or any microbe in the environment may occur via vertical or horizontal gene transfer processes.

Even though there is rapid progress in the advancement of biotechnology, and the production of genetically engineered crops, the same investigative vigour lags behind regarding showcasing the many opportunities for solving current and future global challenges as well as the potential biosafety considerations. As indicated by Lu (2008) gene flow cannot be considered a risk per se, since it is a natural process and part of evolutionary processes. Furthermore, environmental concerns anticipated from transgene escape also depend on whether the gene will be efficiently expressed under *ex-vitro* conditions or not. These are conditions outside the normal laboratory conditions in which the genetic transformation processes of both the bacterium and hosts are conducted. Nonetheless, based on available biological knowledge, transgenes are well controlled and the possibility of gene escapes is effectively reduced. Stricter regulations and legislation, including measures such as confinements and transgene flow containment and mitigation methodologies, are available to reduce the impacts of transgene escape into the environment.

12.11 ABBREVIATIONS

ALS Acetolactate synthase
5'AMP Adenosine-5-monophosphate
DNA Deoxyribonucleic acid
GEM Genetically engineered microorganisms
GM-crops Genetically modified crops
GMO Genetically modified organisms
Goi Gene of interest
HGF Horizontal gene flow
QTL Quantitative trait loci
RNA Ribonucleic acid
T-DNA Transferred DNA
VGF Vertical gene flow
Vir Virulence

REFERENCES

Balakireva AV and Zamyatnin AA. (2016). Properties of gluten intolerance: Gluten structure, evolution, pathogenicity and detoxification capabilities. *Nutrients* 8(644), 1–27.

Barampuram S and Zhang ZJ. (2011). Recent advances in plant transformation. In Birchler J (Eds) *Plant Chromosomes Engineering. Methods in Molecular Biology (Methods and Protocls).* Humana Press, Totowa. pp. 1–35.

Bhalla PL and Smith N. (1998). *Agrobacterium tumefaciens*-mediated transformation of cauliflower, *Brassica oleracea* var. botrytis. *Molecular Breeding* 4, 531–541.
Breiteneder H and Radauer C. (2004). A classification of plant food allergens. *Journal of Allergy and Clinical Immunology* 113, 821–830.
Brookes G and Barfoot P. (2015). Environmental impacts of genetically modified (GM) crops use 1996–2013: Impact on pesticide use and carbon emissions. *GM Crops and Food* 6(2), 103–133.
Burmeister AR. (2015). Horizontal gene transfer. *Evolution, Medicine and Public Health* 1, 193–194.
Chateau S, Sangwan RS and Sangwan-Norcel BS. (2000). Competence of *Arabidopsis thaliana* genotypes and mutants for *Agrobacterium tumefaciens*-mediated gene transfer: Role of phytohormones. *Journal of Experimental Botany* 51(352), 1961–1968.
Chaumont F, Bernier B, Buxant R, Williams ME, Levings CS and Boutry M. (1995). Targeting the maize T-urf13 product into tobacco mitochondria confers methomyl sensitivity to mitochondrial respiration. *Proceedings of the National Academy of Sciences of the United States of America* 92(4), 1167–1171.
De Brondt A, Eggermont K, Pennineks I, Goderis I and Broekaert WF (1996). *Agrobacterium*-mediated transformation of apple (*Malus x domestica* Borth): An assessment of factors affecting regeneration of transgenic plants. *Plant Cell Reports* 15, 549–554.
Eapen S and George L. (1994). *Agrobacterium tumefaciens* mediated gene transfer in peanut (*Arachis hypogea* L.) *Plant Cell Reports* 13, 582–586.
Emamalipour M, Seidi K, Vahed SZ, Jahanban-Esfahlan A, Jaymand M, Majdi H, Amoozgar Z, Chitkusher LT, Javaheri T, Jahanban-Esfahlan R and Zare P. (2020). Horizontal gene transfer: From evolutionary flexibility to disease progression. *Frontiers in Cell Developmental Biology* 8(229), 1–16.
Espinoza-Herrera J, Martinez LM, Serba-Saldivar SO and Chuck-Hernandez C. (2021). Methods for the modification and evaluation of cereal proteins for the substitution of wheat gluten in dough systems. *Foods* 10(1), 118.
Fontes EMG, Pires CSS, Sujii ER and Panizzi AR. (2002). The environmental effects of genetically modified crops resistant to insects. *Neotropical Entomology* 31(4), 497–513.
Franklin G and Sita GL. (2003). *Agrobacterium tumefaciens*-mediated transformation of eggplant (*Solanum melongena* L.) using root explants. *Plant Cell Report* 21, 549–554.
Fredrickson AG and Stephanopoulos G. (1981). Microbial Competition. *Science* 213(4511), 972–979.
Fujieda N, Ichihashi H, Yuasa M, Nishikawa Y, Kurisu G, Itoh S. (2020). Cupin variants as a macromolecular ligand library for stereoselective Michael addition of nitroalkanes. *A Journal of the German Chemical Society* 59(20), 7717–7720.
Galeano P, Debat CM, Ruibal F, Fraguas LF and Galvan GA. (2010). Cross-fertilization between genetically modified and non-genetically modified maize crops in Uruguay. *Environmental Biosafety Research* 9(3), 147–154.
Gasparis S, Bregier C, Orczyk W and Nadolska-Orczyk A. (2008). *Agrobacterium*-mediated transformation of oat (*Avena sativa* L.) cultivars via immature embryos and leaf explants. *Plant Cell Reports* 27, 1721–1729.
Gelvin SB. (2003). *Agrobacterium*-mediated plant transformation: The biology behind the "Gene-Jockeying" tool. *Microbiology and Molecular Biology Reviews* 67(1), 16–37.
Hiei Y, Ishida Y and Komari T. (2014). Progress of cereal transformation technology mediated by *Agrobacterium tumefaciens*. *Frontiers in Plant Science* 5(628), 1–11.
Hiei Y, Ohta S, Komari T, Kumashiro T. (1994). Efficient transformation of rice (*Oryza sativa* L.) mediated by *Agrobacterium* and sequence analysis of the boundaries of the T-DNA. *The Plant Journal* 6(2), 271–282.

Ho CK, Chang SH, Tsay JY, Tsai VLC and Chen ZZ. (1998). *Agrobacterium tumefaciens*-mediated transformation of *Eucalyptus camaldulensis* and production of transgenic plants. *Plant Cell Reports* 17, 675–680.

Ignacimuth S and Kannan P. (2013). Agrobacterium-mediated transformation of pearl millet (*Pennisetum typhoides* (L.) R. Br) for fungal resistance. *Asian Journal of Plant Sciences* 12(3), 97–108.

Ishida Y, Saito H, Ohta S, Hiei Y, Komari T and Kumashiro T. (1996). High efficiency transformation of maize (*Zea mays* L.) mediated by *Agrobacterium tumefaciens*. *Nature Biotechnology* 14, 745–750.

Ishida Y, Tsunashima M, Hiei Y and Komari T. (2015). Wheat (*Triticum aestivum* L.) transformation using immature embryos. In Wang K (Eds) *Agrobacterium Protocols. Methods in Molecular Biology*, Springer, New York. pp. 189–198.

Jeyaramrajo RP and Meenakshi SN. (2005). *Agrobacterium tumefaciens*-mediated transformation of embryogenic tissues of tea (*Camellia sinensis* (L.) O. Kuntze). *Plant Molecular Biology Reporter* 23(3), 299–300.

Jones JDG. (1996). Plant disease resistance genes: Structure, function and evolution. *Current Opinion in Biotechnology* 7, 155–160.

Karmakar K, Kundu A, Rizvi A, Dubois E, Severac D, Czernic P, Carkieaux F and DasGupta M. (2019). Transcriptomic analysis with the progress of symbiosis in 'Crack-Entry' legume *Arachis hypogaea* highlights its contrast with 'Infection Thread' adapted legumes. *Molecular Plant-Microbe Interaction* 32(3), 271–285.

Koetle MJ, Finnie JF, Balazs E and van Staden J. (2015). A review on factors affecting the *Agrobacterium*-mediated genetic transformation in ornamental monocotyledonous geophytes. *South African Journal of Botany* 98, 37–44.

Koncz C, Németh K, Rédei GP and Schell J. (1992). T-DNA insertional mutagenesis in *Arabidopsis*. *Plant Molecular Biology* 20, 963–976.

Kramkowska M, Grzelak T and Czyzewska K. (2013). Benefits and risks associated with genetically modified food products. *Annals of Agricultural and Environmental Medicine* 20(3), 413–419.

Krishnamurthy KV, Suhasini K, Sagore AP, Meixner M, Kathen A, Pickardt T and Scheider O. (2000). *Agrobacterium* mediated transformation of chickpea (*Cicer arietinum* L.) embryo axes. *Plant Cell Reports* 19, 235–240.

Ladics GS. (2018). Assessment of the potential allergenicity of genetically-engineered food crops. *Journal of Immunotoxicology* 16(1), 43–53.

Lee TH, Ho KH and Leung TF. (2017). Genetically modified foods and allergy. *Hong Kong Medical Journal* 23(3): 291–295.

Lenski RE. (1993). Evaluating the fate of genetically modified microorganisms in the environment: Are they inherently less fit? *Experientia* 49, 201–209.

Lu B-R. (2008). Transgene escape from GM crops and potential biosafety consequences: An experimental perspective. *Collection of Biosafety Reviews* 4, 66–141.

Luthy P. (1991). Risk assessment of deliberate release of genetically engineered organisms. *Biotechnology and Biotechnological Equipment* 5(3), 43–47.

Mangena P, Mokwala PW and Nikolova RV. (2015). In vitro multiple shoot induction in soybean. *International Journal of Agriculture and Biology* 17, 838–842.

Manickavasagam M, Ganapathi A, Anbazhagan UR, Sudhakar B, Selvaraj N, Vasudevan A and Kasthurirengan S. (2004). *Agrobacterium*-mediated genetic transformation and development of herbicide-resistant sugarcane (*Saccharum* species hybrids) using axillary buds. *Plant Cell Reports* 23, 134–143.

Marthe C, Kumlehn J and Hensel G. (2015). Barley (*Hordeum vulgare* L.) transformation using immature embryos. In Wang K (eds), *Agrobacterium Protocols. Methods in Molecular Biology*, Springer, New York. pp. 71–83.

Matveeva TV and Lutova LA. (2014). Horizontal gene transfer from *Agrobacterium tumefaciens* and related bacterial species. *Frontiers in Plant Science* 5(326), 1–11.

Metz TD, Dixit R and Earle ED. (1995). *Agrobacterium tumefaciens*-mediated transformation of broccoli (*Brassica oleracea* var. italic) and cabbage (*B. oleracea* var. capitate). *Plant Cell Reports* 15, 287–292.

Mills EN, Jenkins J, Marigheto N, Belton PS, Gunning AP and Morris VJ. (2002). Allergens of the cupin superfamily. *Biochemical Society Transactions* 30(Pt 6), 925–929.

Nap J-P, Bijvoet J and Stiekema WJ. (1992). Biosafety of kanamycin-resistant transgenic plants. *Transgenic Research* 1, 239–249.

Newell CA, Brown NJ, Liu Z, Pflug A, Gowik U, Westhoff P and Hibberd JM. (2010). *Agrobacterium tumefaciens*-mediated transformation of *Cleome gynandra* L., a C4 dicotyledon that is closely related to *Arabidopsis thaliana. Journal of Experimental Biology* 61(5), 1311–1319.

Olhoft PM, Lin K, Galbraith J, Nielsen NC and Somers DA. (2001). The role of thiol compounds in increasing *Agrobacterium*-mediated transformation of soybean cotyledonary node cells. *Plant Cell Reports* 20, 731–737.

Opabode JT. (2006). *Agrobacterium*-mediated transformation of plants: Emerging factors that influence efficiency. *Biotechnology and Molecular Biology Review* 1(1), 12–20.

Pepper IL and Gentry TJ. (2015). Microorganisms found in the environment. In Pepper IL, Garbe CP, Gentry TJ. *Environmental Microbiology*, 3rd eds. Academic Press, Waltham. pp. 9–36.

Prakash D, Verma S, Bhatia R and Tiwary BN. (2011). Risks and precautions of genetically modified organisms. *International Scholarly Research Notices* 369573, 1–13.

Raghavendra P and Pullaiah T. (2018). Future of cellular and molecular diagnostics: Bench to bedside. In Raghavendra P, Pullaiah (eds) *Advances in Cell and Molecular Diagnostics.* Academic Press. Cambridge, Massachusetts. pp. 203–270.

Rizwan M, Hussain M, Shimelis H, Hameed MV, Atif RM, Azhar MT, Qamar Z and Asif M. (2019). Gene flow from major genetically modified crops and strategies for containment and mitigation of transgenic escape: A review. *Applied Ecology and Environmental Research* 17(5), 1191-11208.

Rosellini, D. (2011). Selectable marker genes from plants: reliability and potential. *Vitro Cellular and Developmental Biology- Plant* 47, 222–233.

Solis-Ramos LY, Ortiz-Pavon JC, Andrade-Torres A, Porras-Murillo R, Brenes-Angulo A and La Serna EC. (2019). *Agrobacterium tumefaciens*-mediated transformation of common bean (*Phaseolus vulgaris*) var. Brunca. *International Journal of Tropical Biology and Conservation* 67(2), 83–93.

Takken FLW and Joosten MHAJ. (2000). Plant resistance genes. Their structure, function and evolution. *European Journal of Plant Pathology* 106, 699–713.

Tang W. (2001). *Agrobacterium*-mediated transformation and assessment of factors influencing transgene expression in loblolly pine (*Pinus taeda* L.). *Cell Research* 11, 237–243.

Tsuji H, Kimoto M and Natori Y. (2001). Allergens in major crops. *Nutrition Research* 21, 925–934.

Viljoen, C. and Chetty, L. (2011). A case study of GM maize gene flow in South Africa. *Environmental Sciences Europe* 23(8), 1–8.

White FF, Potnis N, Jones JB and Koebnik R. (2009). The type III effectors of *Xanthomonas. Molecular Plant Pathology* 10(6), 749–766.

Wu E, Lenderts B, Glassman K, Berezowska-Kaniewska M, Chriestensen H, Asmus T, Zhen S, Chu U, Cho MJ and Zhao ZY. (2014). Optimised *Agrobacterium*-mediated sorghum transformation protocol and molecular data of transgenic sorghum plants. *In Vitro Cellular and Developmental Biology-Plant* 50(1), 9–18.

Wu J, Zhang X, Nie X and Luo X. (2008). High-efficiency transformation of *Gossypium hirsutum* embryogenic calli mediated by *Agrobacterium tumefaciens* and regeneration of inseed-resistant plants. *Plant Breeding* 124(2), 142–146.

Zhang W-J, Dewey RE, Boss W, Phillippy BQ and Qu R. (2013). Enhanced *Agrobacterium*-mediated transformation efficiencies in monocots cells is associated with attenuated defense responses. *Plant Molecular Biology* 81, 273–286.

Zvereva AS and Pooggin MM. (2012). Silencing and innate immunity in plant defense against viral and non-viral pathogens. *Viruses* 4, 2578–2597.010237470.

13 The Regulatory Management of Transgenic Plants

13.1 INTRODUCTION

Many countries have developed legislation and compliance measures to assess GMO risks to public and environmental health. In most of these countries, transgenic plants such as Roundup Ready (RR) soybeans are regulated under specific GMO acts, which normally place strict compliance measures on the research, production and marketing of GMOs and their products. Since the start of genetic engineering in the United States, the US Food and Drug Administration (FDA), the US Environmental Protection Agency (EPA) and the US Department of Agriculture (USDA) ensure that GMOs are safe for human and animal consumption, together with the environment at large. African countries including South Africa also adhere to the international food code by the Food and Agriculture Organisation (FAO) of the United Nations and the World Health Organization (WHO), including the Cartagena Protocol on Biosafety (CPB). Many frameworks have been developed by individual countries to coordinate and describe how these agencies should work to regulate GMOs.

However, such legislations and regulations contain guidelines for risk assessment that are generally related to the European Union (EU) Commission and European Food Safety Authority (EFSA) protocols whose steps are exemplified in Figure 13.1. EU and EFSA risk assessment procedures significantly differ from the US risk assessment framework by being stricter and less technically harmonised in terms of the steps (Figure 13.1) involved in conducting risk assessments on GM-crops (Hilbeck et al. 2020). The guidelines presented in Figure 13.1 represent some of the approaches used to objectively identify and evaluate potential adverse risks originating from GMOs on a case-by-case basis. However, the implementation and interpretation of some of these regulations remain controversial and highly problematic for technology developers and food producers in some countries that have postured as anti-GMO. Such countries will be highlighted and briefly reviewed later in this chapter.

Many scientists continue to investigate more approaches and techniques for the efficient genetic transformation of plants, particularly for recalcitrant species, like soybean. In theory and practice, as it was initially collectively reported by Herbert W. Boyer, Stanley N. Cohen and Paul Berg that genetic materials of one organism can be artificially expressed in another organism to achieve DNA recombination, many critics are still reluctant to understand the harmlessness and dynamics factors involved in this process. Lack of understanding may lead to different analysis,

DOI: 10.1201/b22829-13

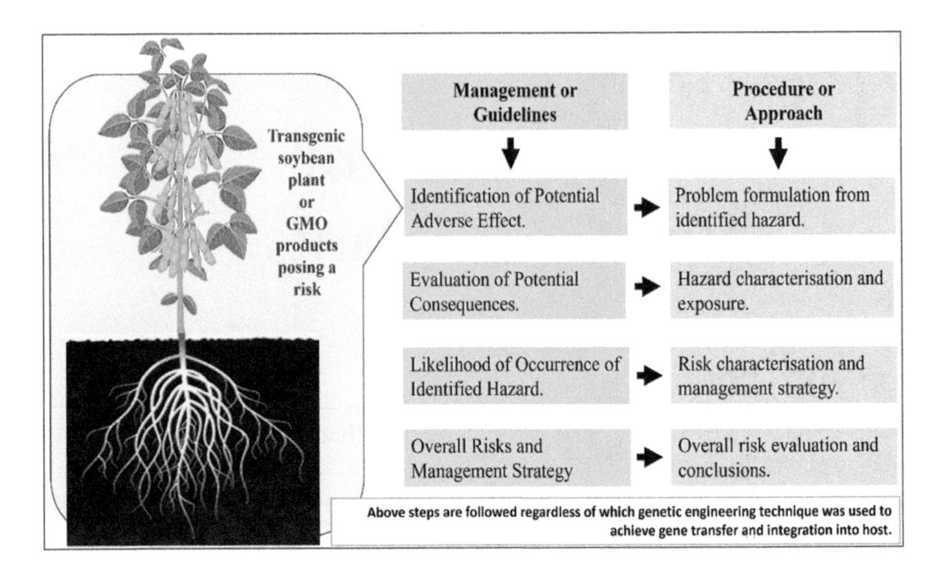

FIGURE 13.1 Summary of the general steps informing framework guidelines or approaches used in the development of GMO risk assessment. This set of generic guidelines emanates from the document developed by the European Food Safety Authority's GMO panel (Hilbeck et al. 2020).

contradictory judgements and wrongful safety or potential risk assessment by different role players. While bearing all of these issues in mind, this chapter will then describe some of the basic regulatory requirements affecting the development and domestication or cultivation of GM-crops, including the application of Roundup Ready-soybeans in agriculture.

13.2 GLOBAL APPROACH TO GMO REGULATIONS

GMO popularly refers to animals, microorganisms and plants that have their genetic material altered through recombinant DNA technology. The regulation of GMOs is necessary, particularly for products consumed as foods. Regulatory processes are important in making sure that the food we eat is safe and of high standards. For many agencies that have the responsibility to protect human and environmental health, regulation should be a more detailed oversight process from inception all the way until the GM-crops reach the consumer. Recommendations and suggestions were made that, while GM-crop regulations must be enforced since they are critically important for protecting consumer health and the environment, legislators or law makers must also bear in mind that these GM-crops such as RR-soybeans play a key role in harnessing the benefits of genetic engineering technology. Transgenic plants should be regulated efficiently, without their economic and social benefits concealing their potential risks.

Although many countries agree with GMO regulation, there are, however, global disagreements with the way that these transgenic products are regulated. Arguments continue while some other developing nations still lack the capacity and resources to adequately regulate GMOs. In 1975 when Cohen reported the invention and creation of novel varieties by transferring genes between species or even from foreign sources, regulation was immediately enforced by the US state government. Enforcement was done due to the anticipation that scientists and researchers would use the invention to maximise profits, attempting to achieve larger economic gains (Hilbeck et al. 2020), often at the expense of consumers and the poor. Currently, there is a great need for consensus among countries on the harmonisation of regulatory systems across the globe as earlier recommended by FAO (2003). The existing regulatory frameworks are different as they were developed by individual countries for their own domestic regulatory purposes and needs. It suffices to say that existing international guidelines such as the Codex and CPB can assist in formulating such a global regulatory framework.

13.3 RISK ANALYSIS AND MANAGEMENT

The cultivation of genetically modified crops has increased since the first GMO was introduced in 1996. Currently, the global cultivation of genetically altered crops has reached a total of 190 million hectares, of which soybean make up 50% of the biotech cultivated crops followed by 30% maize, 13% cotton and only 5% canola (Turnbull et al. 2021). It is suggested that approximately 17 million farmers worldwide grow GM-crops and many of them are found in developing countries. Like many other foods, increased cultivation of GM-crops implies that regulation strategies must also be continuously improved. A systematic approach that involves identifying, analysing and controlling any risks associated with GMO cultivation must be continuously conducted. As observed in many other industrial processes, the purpose of this exercise is mainly to mitigate the impact of potential hazards. Some critical aspects or components of risk analysis are briefly summarised below for potential GM-crop cultivation and release into the outside environment..

13.3.1 Risk Assessment of GM-Crops

The risk analysis of GM-crops involves a systematic evaluation aimed at identifying and analysing any potential risks or issues that may arise as a result of the cultivation and release of a particular GM-crop. Any characteristic traits of the GMO that may be detrimental must be identified through the risk assessment step. This must be an ongoing process involving GMO testing which should also be continuously updated and repeated whenever necessary.

13.3.2 Risk Communication of GM-Crop Analysis

The process of risk communication mainly relies on risk assessors or regulators, together with other legislators who work on controlling breeding technologies. When assessment is being done, whether by one or more agencies, the people involved must be able to exchange information and opinions on any potential risks. Most importantly, that information must be shared with concerned parties, particularly the country's administrative leadership and the public. Both risk assessment and communication may lead to the promotion or banning of the GM-crop being tested. For instance, GM-crops were not allowed in Korea, partly as a result of risk analysis studies conducted in that country. The genetically modified soybeans that were used in Korea were being imported from other countries for food and forage use only (Kim et al. 2015).

13.3.3 Risk Management of Potential Transgenic Hazards

Another component of risk analysis is risk management. This is a more proactive control and evaluation of risks associated with the development and adoption of genetically altered crops. For safety management of environmental risks and assessment, this step usually includes separation of field trials, efficient management of field facilities, proper cultivation practices that prevent cross-pollination and proper management of data gathered from such risk analysis studies. But of paramount importance to this stage is the safety management strategies that ensure the prevention of unintended gene flow by seeds or pollen dispersal, including deterrence of gene escapes by soil runoffs.

13.4 REGULATION AND SANCTIONS OF TRANSGENIC PRODUCTS

There are countries that have imposed bans on the cultivation and consumption of GM-crops despite the existence of stricter regulations that are already in place to monitor, control and approve GMO products. Table 13.1 illustrates some of the countries that strongly oppose the application of genetic engineering. Most of these countries are found in the European Union (EU). France, for instance, banned the growing of GM-crops because of both public hostility and the legal restrictions even though the country imports substantial amounts of GMO products from abroad (Kuntz 2014). The controversies and uninformed reporting by private/public broadcast or newspaper publications criticising GMOs reduced the support for agricultural biotechnology by the French government, the Middle East and some African countries. The nature, properties and applications of genetic engineering will not inherently allow the implementation of this gene technology without proper controls, risk assessment and management.

Already, stringent conditions are required before the gene of interest is isolated, transferred and integrated into the genome of the host organisms. Whenever these conditions are not adequately met, the efficiency and frequency of genetic transformation will drop, leading to the loss of the transgene during this process. Thus, the release of GM products under confined environmental conditions goes through vigorous risk assessments and testing. GMOs will not under any circumstances be released if human and environmental health will be compromised. Such demonstrations also include, for example, the biosafety and implementations of regulations like in the case of kanamycin-resistant transgenic plants, as well as the introduction of any desirable genes for human consumption.

TABLE 13.1
2018/2019 production stats of countries permitting or imposed a ban on the cultivation of genetically engineered crops (Mangena 2020)

Crop:	Country	Yield (mmt)	Area (m.ha)	Production (mmt.ha^{1})	GM crops
Cotton:	Brazil	12.8	1.6	0.00175	++
	China	27.8	3.5	0.00175	+
Corn:	Brazil	100.0	17.5	5.18	++
	Russia	77.0	57.8	2.64	–
	Algeria	4.0	19.2	1.74	–
	Tunisia	1.5	0.6	1.72	+
	Morocco	2.9	0.1	2.19	+
	Syria	4.8	1.6	1.45	+
	Iraq	4.8	2.4	2.72	+
	Iran	16.8	6.7	1.81	–
	India	100.0	29.8	2.28	+
Rice:	China	7.03	30.2	148.5	+
	India	3.91	44.5	116.0	+
	Indonesia	4.79	12.2	37.1	–
	Brazil	6.29	1.75	7.48	++
	Egypt	8.78	0.46	2.8	+
	United States	1.18	8.62	7.12	++
Soybean:	United States	123.7	35.7	3147.9	++
	Brazil	3.24	36.1	117.0	++
	China	1.89	8.4	15.9	+
	India	1.05	11.0	11.5	+
	Canada	1.47	2.74	4.03	+
	Russia	1.47	2.74	4.03	–
	Indonesia	1.27	0.41	0.52	–
	Serbia	2.84	0.22	0.63	–
	Mexico	1.25	0.19	0.34	+
	South Africa	1.75	0.73	1.28	+
	Iran	2.29	0.07	0.16	–

Note: mmt – million metric tons, m.ha – million hectares, and mmt.ha^{-1} – million metric tons per hectare. GM-crops – countries that allow cultivation of genetically engineered crops (+), expanded the area cultivated with genetically engineered crops (++), restrictions and ban on import, distribution and commercialisation of genetically modified crops (–).

Proven cases of transgenic cotton and canola in countries like India and Australia, respectively, show that these genetically improved crops serve as an effective measure to mitigate current and future global challenges (Raman 2017). As such, even if the number of important crops and organisms that are amenable to genetic manipulation is rapidly increasing, these developments are currently hampered, not by the transformation protocol barrier only, but also by legislative regulations that are already in place. Such strict regulations are often established without sufficient scientific insights and evidence that strongly support concerns emanating from the release and use of transgenic products. The basis of all the concerns or potential challenges that are frequently raised also remains controversial and speculative. These remain speculations because most of these expected consequential effects are not yet fully known and will remain unknown for several years to come since no empirical evidence has emerged in support of those claims.

As many reports show, regulations may be developed based on possible or potential GMO effects, such as the antibiotic-resistant transgene assumed to render both the crop and wild plant relatives of the transgenic crop herbicide resistant. Apparently in this case, risk assessment judgements will favour a ban of such a transgenic crop, rendering it unsuitable for human/animal consumption and industrial processing (Nap et al. 1992). However, increases in the number of concerns raised regarding the use of transgenic materials will contribute to the justified legislations supporting the banning of GMO products regardless of whether risk analysis is conducted objectively and in a justified manner within the obligations of their countries' constitutions. Therefore, all of this indicates that the coordination and harmonisation of GMO regulations across the globe is of particular importance, especially for developing countries the majority of which still lack regulatory systems to deal with transgenics.

13.5 SUMMARY

Bioethics and biosafety in gene transfer or regulatory requirements are critical in safeguarding the wellbeing of humans, animals and the environment. Legal frameworks have been set up in many countries to ensure the safe development of this technology, especially those paying special attention to human health and the environment. Legislation and regulations are important to guard against the possible environmental health and safety concerns that may cause pollution in the surroundings, foods, soil and water, including the conservation of the natural genetic resources. Critics of biotechnology have been speculating that transgenic products poison the environment and cause genetic pollution. They believe that these result in the spread of antibiotic-resistant genes or 'super weeds'.

In reviews providing technological and agroeconomic prospects regarding the use of biotechnology in agriculture, May et al. (2005), Pidgeon et al. (2007), Kleter et al. (2008), Heap and Duke (2018) and Duke (2018) sufficiently highlighted the impact and safety of biotechnological products such as GM-crops on the environment and animal health. Many papers took the opportunity to discuss concerns and safety issues raised by different groups of people based on their uncertainties and insecurities. However, dissatisfied lobby groups continue to predict potential adverse effects that might be caused to the environment and health and safety, associated with the applications or consumption of genetically altered products.

For instance, critics claimed that glyphosate-resistant transgenic crops altered mineral nutrition through glyphosate's ability to chelate divalent metal cations.

However, the resistance of the glyphosate-resistant crops indicates that chelating metal cations do not contribute to the herbicidal activity or do not have any significant effects on mineral nutrition (Duke 2018). Finally, regulation and interrogation of safety issues are undoubtedly warranted. But emphasis must also be placed objectively on the insights that may validate or demystify some of the myths surrounding this technology and to promote its applications in various industries, especially in a more sustainable, safe and environmentally friendly manner.

13.6 ABBREVIATIONS

CPB	Cartagena Protocol on Biosafety
EFSA	European Food Safety Authority
EU	European Union
FAO	Food and Agriculture Organisation of the United Nations
GM-crops	Genetically modified crops
GMO	Genetically modified organisms
USDA	United States Department of Agriculture
USEPA	United States Environmental Protection Agency
USFDA	United States Food and Drug Administration
RR-soybeans	Roundup Ready-soybean
WHO	World Health Organization

REFERENCES

Duke SD. (2018). The history and current status of glyphosate. *Pest Management Science* 74(5), 1027–1034.

Food and Agriculture Organisation (FAO) of the United Nations. (2003). Regulating GMOs in developing and transition countries. In Summary Document to Conference 9 of the FAO Biotechnology Forum, Rome, Italy. Date: 28 April to 1 June 2003.

Heap I and Duke SD. (2018). Overview of glyphosate-resistant weeds worldwide. *Pest Management Science* 74(5), 1040–1049.

Hilbeck A, Meyer H, Wynne B and Millstone E. (2020). GMO regulations and their interpretation: 'How EFSA's guidance on risk assessments of GMOs is bound to fail. *Environmental Sciences Europe* 32(54), 1–15.

Kim HJ, An J-H and Han T-H. (2015). Study on the safety management of environmental risk assessment field trial of genetically modified crops in Korea. *Agricultural Science and Technology Research* 50, 9–14.

Kleter GA, Harris C, Stephenson G and Unsworth J. (2008). Comparison of herbicide regimes and the associated potential environmental effects of glyphosate-resistant crops versus what they replace in Europe. *Pest Management Science* 64(4), 479–488.

Kuntz M. (2014). The GMO case in France: Politics, lawlessness and postmodernism. *GM Crops and Food* 5(3), 163–169.

Mangena P. (2020). Genetic transformation to confer drought stress tolerance in soybean (*Glycine max* L.). In Guleria P, Kumar V and Lichtfouse E (eds), *Sustainable Agriculture Reviews. Sustainable Agriculture Reviews*, vol 45. Springer, Cham. pp. 193–224.

May MJ, Champion GT, Dewar AM, Qi A and Pidgeon JD. (2005). Management of genetically modified herbicide-tolerant sugar beet for spring and autumn environmental benefits. *Proceedings of the Royal Society-B* 272, 111–119.

Nap J-P, Bijvoet J and Stiekema WJ. (1992). Biosafety of kanamycin-resistant transgenic plants. *Transgenic Research* 1, 239–249.

Pidgeon JD, May MJ, Perry JN and Poppy GM. (2007). Mitigation of indirect environmental effects of GM crops. *Proceedings of the Royal Society-B* 274, 1475–1479.

Raman R. (2017). The impact of genetically modified (GM) crops in modern agriculture: A review. *GM Crops and Food* 8(4), 195–208.

Turnbull C, Lillemo M and Huoslef-Eide TAK. (2021). Global regulation of genetically modified crops amid the gene edited crop boom: A review. *Frontiers in Plant Science* 12(630396), 1–19.

14 Current and Future Prospects in Genetic Engineering

14.1 INTRODUCTION

The extraordinary capacity of *Agrobacterium tumefaciens* to serve as a vehicle for translocating foreign genetic materials between different organisms made recombinant DNA technology seem more possible. When Smith first discovered this bacterium in 1892, little did he know that this would in many ways bring about surprising observations. This is made possible from its unusual nutritional requirements needed for pathogenesis, the role of its type IV secretion system to form a T-complex in the host and the transfer and integration of nucleotide sequences found between border sequences of T-DNA into host cells. *Agrobacterium* made plant transformation a truly remarkable process, and it is still currently being continuously standardised for further applications in different food and feed crops. Plant transformation has led to the genetic improvement of several major monocot and dicot crops. These include the improvement of corn, canola, wheat, alfalfa and soybean that are not easily achievable through conventional breeding (Table 14.1).

Table 14.1 identifies different transgenic crops used as foods for human consumption and feeds for animal consumption throughout the world. Transgenic soybean varieties were first cultivated commercially amid controversies in not less than eight countries. Those countries include Argentina, Bolivia, Brazil, Canada, Paraguay, South Africa, the United States and Uruguay.

This crop presents similar trends to corn, and both of them have managed to reach over 70% global production in 2018. In the United States more than 90% of soybean production is comprised of genetically modified cultivars (Ates and Bukowski 2021). Most of these cultivated GM-crops are intended to improve growth and yield properties of the plant. Therefore, this chapter discusses the status and future prospects of genetically modified crops, including some other oilseed crops, with a particular emphasis on bioengineered soybean, and its impact on agriculture and food security.

14.2 CURRENT CHALLENGES IN PLANT TRANSFORMATION

As discussed in previous chapters, major challenges facing soybean transformation involve recalcitrance, inefficient culture conditions and poor transformation frequencies. However, one-third of the world's edible oils and two-thirds of protein meals are obtained from these bioengineered crops. Routine protocols still need to be

DOI: 10.1201/b22829-14

TABLE 14.1
Current list of world's genetically modified crops published by the United States Department of Agriculture, regulated by the Agricultural Marketing Service (USDA-AMS)

Bioengineered crop	Modified trait(s)	Country of production
Alfalfa	Herbicide tolerance Low lignin	Canada, Mexico, United States
Apple	Non-browning	Canada, United States
Canola	Herbicide tolerance High laurate Reduced phytate Pollination control Male sterility	Australia, Canada, United States
Corn	Insect resistance Herbicide tolerance Increased ear biomass Alpha amylase Increased lysine Pollination control Male sterility	Argentina, Brazil, Canada, United States, Paraguay, Philippines, Uruguay, South Africa, Egypt, Colombia, Chile, European Union, Honduras
Cotton	Herbicide tolerance Insect resistance	Burma, China, India, Brazil, Pakistan, Swaziland, Australia, Costa Rica, South Africa, Unites States, Mexico, Burkina Faso, Argentina, Paraguay, Sudan, Bangladesh, Canada,
Eggplant	Insect resistance	Bangladesh
Papaya	Virus resistance	Canada, China, United States
Pineapple	Increased carotenoids Inhibition of flowering	Costa Rica
Potato	Reduced black spot bruising Late blight protection Reduced free asparagine Decreased reducing sugars Virus resistance	Canada, United States
Salmon	Growth rate	Canada, United States
Squash	Virus resistance	Canada, United States
Sugarbeet	Herbicide tolerance	Canada, United States
Soybean	Herbicide tolerance Insect resistance Altered growth and oil properties	Argentina, Canada, United States, Brazil, Costa Rica, Uruguay, South Africa, Bolivia, Mexico, Paraguay

developed in order to accomplish higher yield, better seed quality, enhanced growth properties and stress resistance. These improvements are potentially achievable using this technology. However, since the first DNA transfer using *A. tumefaciens*, different transformation methods have been used to produce herbicide-tolerant, insect-tolerant and virus-resistant varieties bringing genetically improved crops to the market. Although research clearly shows that most of these traits were difficult to achieve using conventional breeding, this does not imply that genetic transformation can be accomplished in a highly efficient manner.

There are still countless ongoing genomic and tissue culture-based studies intended to establish rapid and efficient *in vitro* regeneration protocols for routine soybean and other legume crop improvements. Generally, there is a rapidly increasing number of functional genomic studies involving *Agrobacterium* and host plant analyses to accelerate genetic transformation efficiencies and associated *in vitro* culture establishment (Anjanappa and Gruissem 2021). The criteria for verifying plant transformation events, culture conditions for the system, role of plant tissue culture (PTC) and transgenic expression strategies are required. As various studies indicate, a major challenge facing PTC-based transformation is the development of an efficiently rapid protocol and vector construct that produces a high proportion of putative transgenic plants. Such developments will be most welcomed as they are needed to make crops more resilient to climate change and ensure food security for the ever-growing world population.

14.3 ROLE OF GM-CROPS IN AGRICULTURE

While plant transformation remains a challenge due to factors mentioned above, including genotype specificity, selection criteria and poor recovery of transgenic plantlets, genetically engineered crops continue to be grown throughout the world for the production of food, feed and medicine. At present, there are about 13 approved GM-crops that are grown in more than 40 countries (English and Kayleen 2020). These crops represent a large portion of global crop production with major species such as cotton, maize, sugarcane and soybean making up more than 75% of cultivated GMOs in those countries. The most predominant traits for these crops are herbicide and insect resistance. Newly introduced varieties include the rollout of the new golden rice and soybean exhibiting enhanced growth rates cultivated in the Philippines and Canada, respectively.

The United States, Nigeria and Kenya are also exploring the rollout of genetically engineered soybeans with increased growth properties, whereby the two African countries only started growing GM-crops recently. The market for transgenic crops in agriculture gained billions of dollars in 2020. The global market for GM-crops recorded an estimated value of US$28 billion, and it is projected to reach over US$45 billion by the year 2027 (Anjanappa and Guissem 2021). From these records, soybean trade nearly made up a quarter or more of the value of global GM-crop production. The value of soybean remains one of the major crops because it serves as a major global export mainly for its oil and protein meal. Soybean serves as the only vegetable crop that contains nine essential amino acids and has become the most important source of proteins for animal and human consumption (Voora et al. 2020). Soy oil is the second most consumed oil after palm oil in the world.

Apart from its nutritional benefits, the cultivation of soybean is concentrated in a few countries (the United States, Brazil and Argentina) which account for more than 80% of the total global soybean production. Increases in productivity were experienced due to the introduction of herbicide-tolerant varieties. These genetically modified varieties are responsible for the crop's rapid expansion among urban and rural farmers, particularly the small-holder farmers. It is now clear that in some countries like Mexico, Brazil and India, soybean production will soon be driving the agricultural sector, overtaking major monocotyledonous crops such as wheat and corn. The adoption of those transgenic varieties mainly for agricultural purposes significantly changed the patterns of trade and consumption, and sparked obstinate disagreements within the agricultural business, politicians, law makers and advocacy groups, as well as the general public.

14.4 THE IMPORTANCE OF GMOs FOR FOOD SECURITY

Food security continues to be a lurking global challenge requiring serious attention from government administrators, legislators, the private sector, civil society and the whole population. Fighting poverty and food insecurity is the responsibility of every living human being since these problems bring uncertainty and may negatively impact a country's sovereignty. Poverty and starvation have the ability to negatively influence political stability, economic growth and population well-being that determine variations or improvements in life expectancy. Poverty exacerbates the spread of diseases and causes increases in crime levels. These and other population characteristics not mentioned above may influence the longevity and health of an entire population (Arora et al. 2016). On the other hand, farmers, plant breeders and plant scientists also have a major role to play in alleviating poverty and ensuring food security.

Genetic engineering tools such as plant transformation, particle bombardment or mutation breeding are of great interest in ameliorating threats brought by food insecurity. As the introduction of GMO technology was used by biotech business to change global market value chains and patterns of trade, similarly, these approaches must be used to influence strategies that will circumvent the effects of poverty and food insecurity. Food insecurity is more rife in developing countries where social stratification reflects deeply divided societies with a hugely unequal distribution of goods, services, income and wealth (McLeod and Nonnemaker 1999). Reports indicate that Brazil adopted GMO technology after 2005, later than countries such as Argentina and China. But Brazil has seen rapid growth in its agricultural value chain following the start of expanded commercial GM-crop cultivation in 2006. The adoption of GM-technology significantly changed trade patterns leading to massive market gains since this country became one of the leaders in supplying high-income countries with the first commercial release of GM-soybeans (Oliveira et al. 2020).

While the adoption of GMOs is still opposed in some countries, application of this technology continues to grow in other countries, particularly in sub-Saharan Africa.

Perhaps the importance of GM-crop cultivation is now being realised due to tremendous increases in growth-inhibiting factors that threaten food security throughout the world. Increases in scorching temperatures, drought, salinity stress and mineral deficiencies result in decreased crop productivity. These stress conditions create irreversible damage to agriculture in developing countries where there is a vast lack of irrigation systems and other inputs. The world is considered to be at a critical junction when it comes to the state of food security and nutrition. The commitments on ending hunger, poverty and all forms of malnutrition can be realised if appropriate agricultural production technologies such as *Agrobacterium tumefaciens*-mediated genetic transformation are enhanced and their products effectively adopted.

14.5 OTHER PLANT BREEDING METHODS

Other plant breeding methods that play a significant role in agricultural developments include mutation breeding, microprojectile bombardment and gene editing. Nevertheless, genomic editing tools have been predominantly used in the modification of the human genome for treating genetic and acquired diseases (Li et al. 2020). In plants, this technique has been used for studying gene expression and functional genomics for counteracting the effects of biotic and abiotic stresses. Stresses such as drought, pests, diseases and weeds serve as major growth-limiting factors to crop growth and productivity. In the past, these stresses were more detrimental and persistent due to limitations associated with conventional breeding. Their unembellished effects in agriculture were mostly realised at that period prior to the introduction of useful biotechnological strategies. In fact, the stresses had uncontrollable effects pre-genetic engineering era, and only lasted until the emergence of the current and continuously explored molecular techniques that serve as new breeding tools.

As previously indicated, these tools are used by researchers to develop varieties with high stress-resistance traits. Plant transformation, for instance, provides options propelling high economic gains by preventing yield losses due to biotic and abiotic stress. This tool also enhances the quality and quantity of the produced foods. About 14–25% and 10–16% of yield can be lost respectively to pathogens and pests every harvest period globally (Tohidfar and Khosravi 2015). Previously, such losses would be difficult to circumvent because only individual species of the same taxon showing a resistant trait could be crossed to produce stress-resistant hybrids. However, if resistance genes did not exist within the gene pool then breeders would usually fail to establish newly improved varieties carrying the desired traits. This clearly indicated that it is necessary to search for alternative new sources of resistant genes capable of being transferred to other crop species. The discussions below highlight the future of biotechnology and the role that some of these alternative techniques can play in agriculture.

14.5.1 Mutation Breeding and Plant Mutagenesis

Induced mutagenesis is one of the most efficient approaches used to acquire new genetic diversity. This tool plays a significant role in achieving genetic variability for desirable quality and stress-resistance traits in crops. This approach has been utilised to extensively create genetic variations or to identify regulatory genes responsible for agroeconomically important traits required in crop improvement (Chaudhary et al. 2019). Mutations can be achieved using chemicals, physical induction or insertional mutagenic treatments in association with whole-genome sequencing to provide an efficient plant improvement protocol. In soybean, a unique radiation facility established in Korea was used to develop commercially viable mutants from 1960. To date, at least five soybean mutants exhibiting early maturity, high yield and seed coat colour change have been developed.

These soybean mutants, together with those genotypes showing other traits like altered null lipoxygenase enzymes, Kunitz trypsin inhibitor and altered protein activity were reported to play a key role in providing additional genetic resources for larger breeding programmes (Ha et al. 2014). All of the induced mutation can be traced back using genomic analysis, harnessing these properties and increasing crop diversity. Furthermore, biological agents such as T-DNA, transposons and CRISPR-Cas9 transgenic technologies were also reported as promising mutation-inducing methods by Espina et al. (2018). Soybean mutants produced are often thoroughly checked for heritability of beneficial traits into new offspring, and the hybrids may be exploited for future breeding programmes as novel sources of germplasm.

14.5.2 Microprojectile Bombardment

Biolistic gene transfer employs high-velocity microprojectiles to deliver foreign genes to intact cells and tissues as discussed in Chapter 7. This technique has also resulted in a range of successful transformation events in a number of species. As previously indicated, it makes use of cell suspension, callus, immature embryo, mature embryo, meristem, leaf apices, microspore and pollen cultures. In monocots, a drought-resistant transgenic sorghum cultivar P898012 was developed using immature zygotic embryo culture. Sorghum is considered the fifth most cultivated cereal crop species for feed, food and biofuel production in Asia as well as in sub-Saharan Africa. In addition, this annual diploid C_4 grass was engineered for increased biomass and high-value products like syrups (Casas et al. 1993; Silva et al. 2021). Maize is the other important food and feed crop that has a number of biotech traits available on the market.

Tremendous progress has been made in *Agrobacterium tumefaciens*-mediated transformation and biolistic maize transformation in generating commercial traits that are currently on the market (Que et al. 2014). General guidelines for routinely increased transformation efficiency and gene expression are also available for monocots compared to dicot species transformation. Since monocot plants like maize, rice and sorghum became targets for biotechnological innovations in the mid-1990s, there are now more registered transgenic cultivars used for commercial agriculture. Biolistic protocols have been successfully used to augment traditional breeding methods for crop improvement. Microprojectile bombardments also circumvented some of the shortcomings facing *Agrobacterium*-mediated genetic transformation.

The lack of efficient regeneration protocols, for example, is solved by the ability of this technique to target cells, tissues or organs that have high morphogenic potentiality. Most researchers may find this tool very simplistic to determine parameters for the optimisation of DNA delivery to target tissues using microprojectiles. However, the application of the technology in legume transformation widely is mainly prohibited by the difficulties in plant regeneration and its high costs of application.

14.5.3 Genomic Editing

Currently, there is a strong view that modern precision breeding through CRISPR-Cas9 technique is rapidly overtaking tools such as *Agrobacterium*-mediated and biolistic gene transfers. Although debates are still raging on whether genetically modified plants produced using this technology should be classified and labelled as GMOs or not, this approach is considered to be taking over genetic engineering of important fruit crops and animals. However, no commercial varieties have been produced using this technology. But research on a range of gene editing techniques is still ongoing, promising improved crops with boosted flavour, biotic and abiotic stress resistance, and even tackling allergens like gluten, cupin, lectin and Kunitz trypsin inhibitors found in soybeans and other legume crops. In the case of allergenic responses to legumes and other related species, a report by Verma et al. (2017) provided comprehensive details of some of the well-known, identified and characterised allergens.

In medicine, these techniques also promise to cure genetic and inherited diseases such as cancer, cardiovascular conditions and rare sight losses like Stargardt disease. In agriculture, gene editing promises to create plants that produce higher yields, that are more nutritious and that are able to endure extreme stress conditions imposed by drought, pests, etc. Currently, research developments are focussed on testing the potential of CRISPR-Cas9 techniques in establishing genetically engineered plants of apple, banana, grapes, kiwifruit, strawberry, tomato and watermelon (Wang et al. 2019). It is now clearer that genomic editing and other newly emerging technologies continue to revolutionise plant breeding and pave the way to new horizons for crop genetic improvements.

14.6 IMPROVEMENT OF CROP TRAITS

Crop growth and productivity characteristics are complex quantitative traits determined mainly by genetic and environmental factors. Evaluations of soybean growth characteristics are required in order to determine outperforming varieties, which can also be used for the breeding of new lines. Soybean varieties with higher plant height and large numbers of nodes, branches and trifoliate leaves are normally selected. Generally, these vegetative traits directly influence agronomic traits such as the number of flowers, pods, grains and 100-seed weight produced by a particular species or population. Therefore, all growth and yield characteristics are used as direct selection criteria for higher yielding varieties, irrespective of whether the plant is directly or indirectly affected. They are also evaluated, where possible, in association with environmental interactions. As the main focus is on yield, the yield of certain soybean varieties can be significantly reduced when they are cultivated in different environments (Li et al. 2020).

Therefore, it is important to develop new soybean varieties using conventional or modern breeding techniques that suit those environmental conditions. Many studies indicate how yield components are selected and correlated with corresponding genes (Li et al. 2008; Debebe et al. 2014; Dao et al. 2017; Li et al. 2020). Furthermore, the yield of individual plants also depends on the population (Figure 14.1). Yield per plant may gradually decrease when the population or area increases as shown in Figure 14.1. Efforts made, as demonstrated in Figure 14.1, to quantify the relationship between plant population and yield indicate that yield may be increased with the increase in efficient utilisation of growth and productivity promoting factors, including stability of environmental conditions (Rana and Rana 2014). Consequently, crop yield results from interactions among all growth and yield components as influenced by different environmental conditions. Any outcome may be a result of the interactions with the environment which may have significant impacts upon the expression of crop productivity per specific area as reported by Beruski et al. (2020).

14.7 DEVELOPMENT OF STRESS-RESISTANT CROPS

The generation of transgenic plants has proved to be one of the most crucial steps in the development of a sustainable agriculture. For commercial agriculture, it is therefore highly desirable that a high-throughput transformation system is established. Such a system should ensure mass recovery and propagation of transgenic plants with high quality in an elite genetic background (Que et al. 2014). The development of *Agrobacterium*-mediated genetic transformation has made the generation of genetically engineered crops simpler and more reliable for agricultural purposes. Progress in plant transformation has made it possible to introduce modifications and insert transgenes at a specific chromosomal target site to induce resistance to stress. Even though this tool still faces inefficiencies and *in vitro* regeneration problems, it

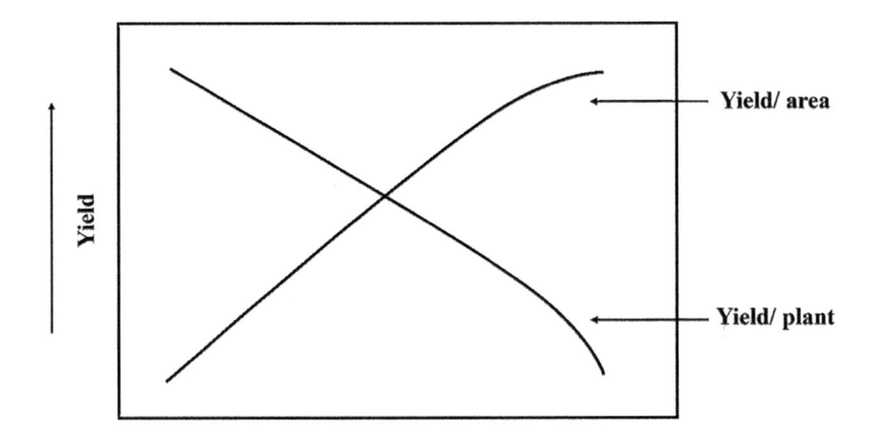

FIGURE 14.1 Prediction of yield performance based on the increase in population or area.

is conceivable that its application in agriculture is yet to bring more positive changes in the food production value chain as presented in Figure 14.2.

Genetic transformation is capable of building value at every stage of agricultural production from the breeding of newly improved lines to harvesting, processing, distribution and consumption (Figure 14.2). The establishment of highly efficient and routine transgenic systems will deliver higher value of agricultural produce and increase profits for each stakeholder from harvesting to consumption. Although research is continuing on the genetic improvement of crops, some countries always experience pressures from numerous production challenges that include the lack of high-yielding cultivars, the lack of stress-resistant varieties and the effects of climate change. A lack of drought-tolerant cultivars has already exerted immense pressure on soybean supply chain, weakened the demand, imposed price increases and generally caused a decline in the availability of this crop for the manufacturing of agricultural raw materials (Mangena 2022).

14.8 LEGISLATION AND SOCIAL ACCEPTANCE

Climate change and food insecurity influence the formation of some of the major legislations developed by different countries across the globe. These legislations are utilised to determine the structure and operation of key stages in the agricultural value chain (Figure 14.2). Generally, supply chains must be regularly improved if the agricultural sector is to meet farmers' needs, consumer demands and the establishment of alignment between genetic improvements and crop productivity. However, there are certain policies governing the seed system, agricultural expansion, processing industries and consumer protection that have a strong emphasis on the prohibition of genetic engineering. These policies pay special attention and focus to issues relating to consumer and environmental risks, without considering the capacity of current and available seed systems. In fact, seed systems must be proactively linked with modern breeding programmes in order to effectively conserve and manage genetic resources, especially those that are required for improving crop diversity.

Additionally, available germplasms must be used by breeders and biotechnologists to develop newly improved varieties that benefit farmers and consumers (Mangena 2022). Even though the use of such genetic stocks, for instance to generate GM crops, has already raised serious criticisms due to the potential impact that GMOs have on human health and the environment, GM-crops, including transgenic soybeans, serve as an abundant and crucial source of feed and nutrition for animal and human consumption. Soybeans, like most of the legumes, provide relatively a high quantity and quality of proteins and essential amino acids (Jacobs et al. 2016). However, these genetically modified cultivars have been restricted and demonised, particularly because of their strong association with pesticides. They are banned in most countries as shown in Chapter 13 irrespective of the many people that use pesticides around their homes, or on their skins as insect repellents.

As such, appropriate legislation and consumer acceptance towards GMOs are required in order to strike a balance between myths and scientific evidence on pleiotropic effects causing negative changes in crops with altered genomes. Clarification on

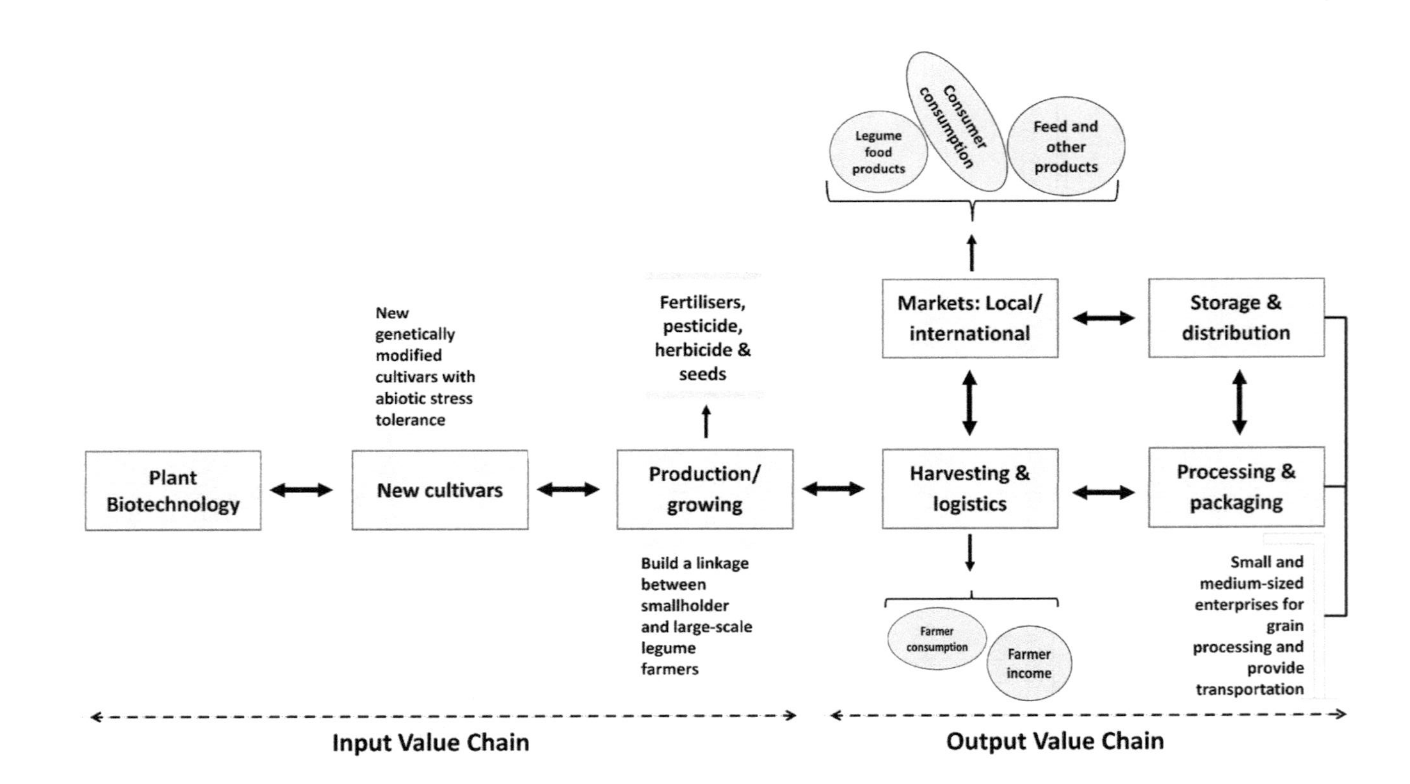

FIGURE 14.2 Illustration of the agricultural value chain involving the cultivation, harvest and productivity of genetically modified crops.

legislative and consumer perspectives must be given since transgenic plants have been met with huge suspicions and fear regardless of the impact they have on agriculture. Since perhaps more novel hazards relating to recombinant DNA technology gain more traction, various allegations against GMOs that lacked scientific evidence still brought about new legislations and regulations. But these critical publications were then mostly found to lack scientific validity by many reviewers (McHughen 2013). It is important that governments develop legislation to safeguard their populations and the environment against the misuse of genetic resources or derivatives thereof. But this should be done without creating unnecessary anxiety among consumers, and with adequate knowledge about the safety of GM-crops and the scientific evidence that governs the development and use of genetically engineered crops.

14.9 FUTURE PROSPECTS

Soybean transformation has a bright future as progress in this genome engineering technology has already made it possible to make modifications and introduce new genetic materials in many crops' genomes. Since the discovery of *Agrobacterium tumefaciens* in 1892 by Erwin Frank Smith (an American plant pathologist, born in New York in 1854) and the first successful soybean transformation by Maud Hinchee (Monsanto Company, Saint Louis, Michigan, United States) were reported (Hinchee et al. 1988), this technique has made major strides in the development of new cultivars with improved commercially valuable traits. For instance, drought-resistant varieties have been introduced to circumvent the massive reductions in the growth and productivity of many crops. In soybeans, drought serves as one of the major causes of yield losses and has the ability to decrease global harvest production to less than 50% (Mohamed and Latif 2017).

Even though plants can acclimatise and adapt to drought stress through morphological, physiological and biochemical responses, all of these extracellular and intracellular responses remain inadequate when the level of stress is more intense and occurring at frequent rates. The genomic makeup of plants can be changed to enable effective adjustments during the period of stress and to make them drought tolerant. In addition, this expression of resistant genes is considered an important approach to increasing stress resistance since the induced traits may be heritable. Therefore, these and other benefits of genetic engineering demonstrate that this technology is an absolute necessity to enhance agricultural production. Plant transformation demand is increasing amid a myriad of challenges facing the entire food system and threats to food security.

These ongoing challenges somewhat discourage the ambitions of feeding the ever-increasing population. Current and future optimisations of the transformation protocols will potentially lead to effective targeted changes in most plants, promising to accelerate crop improvement by increasing the quality and quantity of yields.

14.10 SUMMARY

This chapter as well as other sections in this book have clearly demonstrated that *Agrobacterium tumefaciens*-mediated genetic transformation of soybean, as well as of many other monocot and dicot plants, is a unique genetic exchange. Advances made have contributed to the understanding of fundamental biological principles that led to the development of an entirely new industry in agricultural biotechnology. *Agrobacterium* revolutionised plant molecular genetics and biotechnology with its tremendous capabilities to translocate genetic materials among a wide range of eukaryotic species (Nester 2015). Certainly, this genetic approach has complemented conventional breeding in the introduction of new genetic variations for agricultural purposes. Although the technique exhibits many features that still deserve much more attention and research, it has so far succeeded in the improvement of various agronomic traits for commercial cultivations.

The genetic transformation of soybean by *Agrobacterium* and particle bombardment was established by several laboratories worldwide. In general, soybean transformation frequencies are reported to be genotype-dependent and highly recalcitrant. This suggests that there is a great need to optimise and adapt transformation protocols that are cultivar independent. While DNA transfer frequency could be efficiently quantified through monitoring the expression of β-glucuronidase and green fluorescent protein gene, this protocol is still far from establishing a routine protocol. But genetic transformation via this bacterium is more preferable and desirable than other direct and indirect uptake of DNA such as electroporation, microprojectile bombardment, sonoporation, laser transfection or *Agrobacterium rhizogenes*-mediated transformation.

Many of the above techniques cannot be applied in many crop laboratories worldwide due to prohibitive costs. For instance, effective intracellular delivery methods like electroporation and sonoporation have been extensively investigated but have still not generated overwhelming support as many of them require specialised and sophisticated laboratories mainly operated by niche researchers. Finally, given the many benefits emanating from using these technologies, many people must learn more about their use in generating transgenic plants. Policy makers, farmers and consumers must understand the mechanisms used by *Agrobacterium* to successfully transfer and express its segment of genetic materials (T-DNA) and see what probable hazards could result from such a natural process. Critics should make efforts to understand the natural mechanisms that *Agrobacterium* uses to consequently attach itself to wounded plant cells. Answers to the question of why this bacterium is highly regarded as a 'natural genetic engineer' may demystify many misconceptions, unreasonable restrictions and public scepticism that this modern breeding technology is faced with.

14.11 ABBREVIATIONS

CRISPR-Cas9 Clustered regularly interspaced short palindromic repeats associated protein 9
DNA Deoxyribonucleic acid
GM-crops Genetically modified crops
GMO Genetically modified organisms
PTC Plant tissue culture
T-DNA Transfer-DNA

US$ United States dollars
USADA-AMS United State Department of Agriculture – Agricultural Marketing Service

REFERENCES

Anjanappa RB and Gruissem W. (2021). Current progress and challenges in crop genetic transformation. *Journal of Plant Physiology* 261(153411), 1–13.

Arora A, Spatz E, Herrin J, Riley C, Roy B, Kell K, Coberley C, Rula E and Krumjolz HM. (2016). Pollution well-being measures help explain geographic disparities in life expectancy at the county level. *Health Affairs* 35(11), 2075–2082.

Ates AM and Bukowski M. (2021). *Oil Crops Outlook: December 2021*. Oil Crops Outlook No. (OCS-21L). Department of Agriculture and Economic Research Service, Washington, D.C. USA. pp. 1–9.

Beruski GC, Schiebelbein LM and Pereira AB. (2020). Maize yield components as affected by plant population, planting date and soil coverings in Brazil. *Agriculture* 10(579), 1–20.

Casas NM, Kononowicz AK, Zehr UB, Tones DT, Axtell JD, Butler LG, Bressan RA and Hasegawa PM. (1993). Transgenic sorghum plants via microprojectile bombardment. *Proceeding of the National Academy of Sciences of the United States of America* 90, 11212–11216.

Chaudhary J, Deshmukh R and Sonah H. (2019). Mutagenesis approaches and their role in crop improvement. *Plants* 8(467), 1–4.

Dao A, Sanou J, Traore USE, Gracen V and Danquah EY. (2017). Selection of drought tolerant maize hybrids using path coefficient analysis and selection index. *Pakistan Journal of Biological Sciences* 20(3), 132–139.

Debebe A, Singh H and Tefera H. (2014). Interrelationship and path coefficient analysis of yield components in F4 progenies of tef (*Eragrostis tef*). *Pakistan Journal of Biological Sciences* 17(1), 92–97.

English C and Kayleen S. (2020). *Where are GMO Crops Grown? GLP Infographics Document the Global Growth of Agricultural Biotechnology Innovation*. Genetic Literature Project, Cincinnati, United States. pp. 1–4.

Espina MJ, Ahmed CMS, Bernardini A, Adeleke E, Yadegari Z, Areli P, Pantalone V and Taheri A. (2018). Development and phenotypic screening of an ethyl methane sulfonate mutant population in soybean. *Frontiers in Plant Science* 9(394), 1–12.

Ha B-K, Lee KJ, Velusamy V, Kim J-B, Kim SH, Ahn J-W, Kang S-K and Kim D-S. (2014). Improvement of soybean through radiation-induced mutation breeding techniques in Korea. *Plant Genetic Resources* 12(S1), 54–57.

Hinchee MAW, Connor-Ward DV, Newell CA, McDonnell RE, Sato SJ, Gasser CS, Fischhoff DA, Re DB, Fraky RT and Horsch RB. (1988). Production of transgenic soybean plants using *Agrobacterium*-mediated DNA transfer. *Bio/Technology* 6, 915–922.

Jacob C, Carrasco B and Schwember AR. (2016). Advances in breeding and biotechnology of legumes crops. *Plant Cell, Tissue and Organ Culture* 127, 561–584.

Li H, Yang Y, Hong W, Huang M, Wu M and Zhao X. (2020). Applications of genome editing technology in the targeted therapy of human diseases: Mechanisms, advances and prospects. *Signal Transduction and Targeted Therapy* 5(1), 1–25.

Li YC, Yu D, Xu R and Gai JY. (2008). Effects of natural selection of several quantitative traits of soybean RIL populations derived from the combination of Peking X7605 and RN-9X7605 under two ecological sites. *Scientia Agricultura Sinica* 41, 1917–1926.

Mangena P. (2022). COVID-19 pandemic and agriculture: Potential impact on legumes and their economic value chain. In Faghih N and Forouharfar A (eds), *Socioeconomic Dynamics of the COVID-19 Crisis: Contributions to Economics.* Springer, Cham. pp. 485–506.

McHughen A. (2013). GM crops and foods, *GM Crops and Food* 4(3), 172–182.

McLeod JD and Nonnemaker JM. (1999). Social stratification and inequality. In Aneshensel CS and Phelan JC (eds), *Handbook of the Sociology of Mental Health. Handbooks of Sociology and Social Research.* Springer, Boston. pp. 321–344.

Mohamed HI and Latif HH. (2017). Improvement of drought tolerance of soybean plants by using methyl jasmonate. *Physiology and Molecular Biology of Plants* 23(3), 545–556.

Nester EW. (2015). Agrobacterium: Nature's genetic engineer. *Frontier in Plant Science* 5(730), 1–16.

Oliveira PRS, da Silveira JMFJ, Magalhaes MM and Souza RF. (2020). International trade in GMOs': Have markets paid premiums on Brazilian soybeans? *Revista de Economia e Sociologia Rural* 58(1), e167573, 1–23.

Que Q, Elumalai S, Li X, Zhong H, Nalapali S, Schweiner M, Fei X, Nuccio M, Kelliher T, Gu W, Chen Z and Chilton MDM. (2014). Maize transformation technology development for commericaial event generation. *Frontiers in Plant Science* 5(379), 1–19.

Rana SS and Rana RS. (2014). *Advances in Crop Growth and Productivity.* Department of Agronomy, CSK Himachal Pradesh Krishi Vishvavidyalaya, Palampur. pp. 1–230.

Silva NT, Thomas JB, Dahlberg J and Rhee SY. (2021). Progress and challenges in sorghum biotechnology, a multipurpose feedback for the bioeconomy. *Journal of Experimental Botany* 73(3), 646–664.

Tohidfar M and Khosravi S. (2015). Transgenic crops with an improved resistance to biotic stresses. A review. *Biotechnology, Agronomy, Society and Environment* 19(1), 62–70.

Verma AK, Kumar S, Das M and Dwivedi PD. (2017). A comprehensive review of legume allergy. *Clinical Reviews in Allergy and Immunology* 45, 30–46.

Voora V, Larrea C and Bermudez S. (2020). *Global Market Report: Soybeans.* The International Institute for Sustainable Development (IISD), pp. 1–20.

Wang T, Zhang H and Zhu H. (2019). CRISPR: Technology is revolutionizing the improvement of tomato and other fruit crops. *Horticulture Research* 6(77), 1–13.

15 Photographic Index of Soybean Transformation Cultures

15.1 INTRODUCTION

This chapter provides an index of photographic illustrations demonstrating stages and processes performed during *in vitro* plant tissue culture-based *Agrobacterium*-mediated genetic transformation in soybean. This is a cotyledonary-based system that involves *in vitro* regeneration, T-DNA expression and the use of selective medium for potential screening of transgenic plantlets. The mechanism and application of this *Agrobacterium* vector included the strain EHA101 with a construct derived from the base vector pTF101.1, a derivative of the pPZP binary vector. This binary vector contains the right and left T-DNA border fragments from a nopaline strain with a broad host origin of replication (pVS1) and a spectinomycin-resistant marker gene (*aadA*) for bacterial selection (Figure 15.1).

The plant selectable marker gene cassette consists of a double 35S cauliflower mosaic virus promoter (CaMV), tobacco etch virus translational enhancer (TEV) and phosphinothricin acetyl transferase (*bar*) gene from *Streptomyces hygroscopicus* that confer resistance to the herbicide phosphinothricin and its (Figure 15.1) derivatives. Furthermore, this vector construct also contains the soybean vegetative storage protein terminator which was cloned at the 3' end of the *bar* gene (Paz et al. 2004; Paz et al. 2006). For additional monitoring of gene expression and protein localisation, Green Fluorescent Protein (GFP) can be used to produce a fluorescent product when expressed in transformed plant tissues. GFP is capable of producing a strong green fluorescence when excited by a blue light without requiring any additional gene products; it is not species specific and occurs either due to ubiquitous cellular components or autocatalysis. Green fluorescence is induced upon transgenic plant tissue illumination with a long-wave ultraviolet (UV) source (Chalfie et al. 1994; Harper and Stewart 2000).

15.2 THE SOYBEAN PLANT

Erect soybean [*Glycine max* (L.) Merrill.] plant is grown under natural environmental conditions (Figure 15.2). Plants were grown for seed multiplication, and for obtaining freshly harvested seeds used in *in vitro* regeneration and genetic transformation experiments. Cultivation was performed at Amaloba Nursery, at the University of Limpopo, South Africa. Soybean growth may reach more than a metre in plant height and produces white or purple flowers. The pods are hairy, green and brown when

DOI: 10.1201/b22829-15

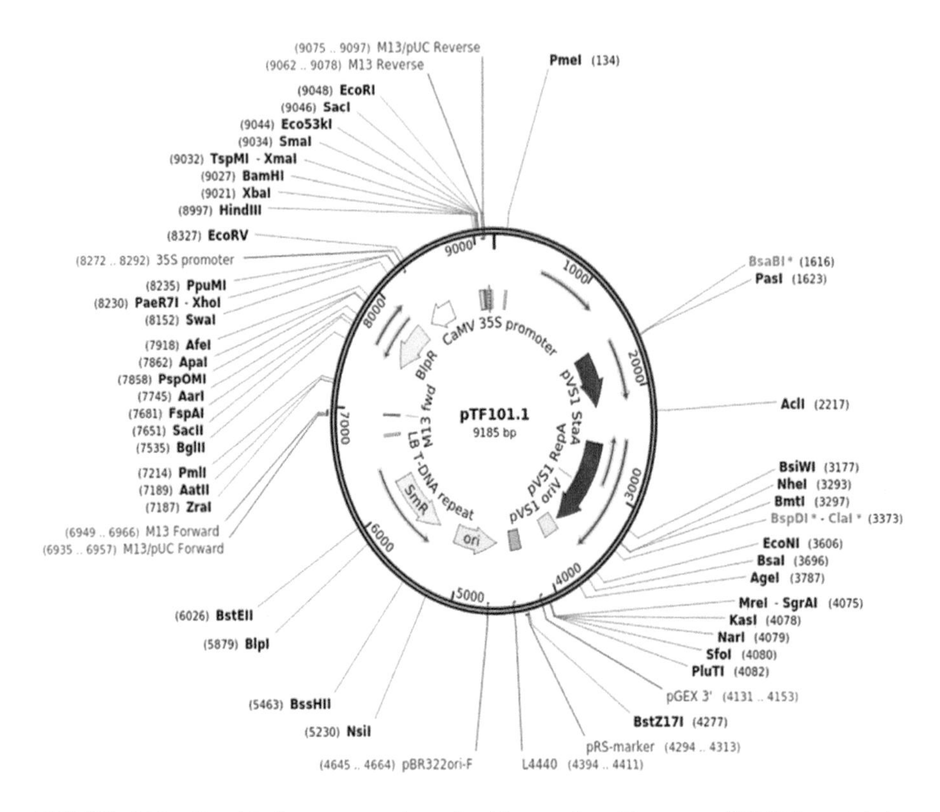

FIGURE 15.1 Graphical representation of a binary plasmid vector pTF101.1 containing double CaMV 35S promoter and tobacco etch virus (TEV) enhancer to drive bialophos resistance gene (*bar*) for plant transformation (Paz et al. 2004).

FIGURE 15.2 The soybean plant.

matured, while seeds can be green, black, brown or predominantly yellow depending on the type of cultivar (Figure 15.2). Seeds of some commercial varieties can be bicoloured, with two to three seeds per pod. Soybeans can be grown in most types of soils but they thrive in warm, fertile, well-drained, sandy loam soils. They are mostly harvested when their leaves have fallen off and seed moisture has dropped to about 13% to improve seed viability and vigour during storage (Mangena 2018) (Figure 15.2).

15.3 SEED STERILISATION AND GERMINATION

All soybean seeds harvested from Amaloba Nursery were disinfected with chlorine gas prior to use in plant tissue culture-based transformation. Normally, for the preparation of both *in vitro* regeneration and *Agrobacterium tumefaciens*-mediated genetic transformation, soybean seeds were first washed with a detergent (domestic liquid soap or dish washer) in tap water to remove soil detritus, and then rinsed a few times with distilled water (Mangena et al. 2015). However, it is very important to note that this sterilisation procedure does not eliminate or prevent any internal contamination that may be contained within the seeds.

The prevention of internal contaminants can be achieved by including antibiotics in a culture medium. Some studies (de Oliveira and Costa 2010; Carey et al. 2015) indicated that the inclusion of antibiotics in regeneration cultures normally hinders cell proliferation for the induction of multiple shoots and subsequent plant establishments (Figures 15.3 and 15.4).

15.4 COTYLEDONARY NODE EXPLANTS

Cotyledonary node explants (single and double coty-nodes) are derived from *in vitro* or *in vivo* germinated soybean seeds. These are the most commonly used type of

FIGURE 15.3 Surface gas sterilisation.

FIGURE 15.4 *In vitro* germination.

FIGURE 15.5 Cotyledonary node explants.

explants in soybean transformation, often with or without pre-existing meristematic cells (Opabobe 2006; Paz et al. 2006; Zhang et al. 2014; Li et al. 2017; Mangena 2019; Bhajan et al. 2019; Mangena 2021). Double coty-nodes are prepared by aseptically excising both the epicotyl and hypocotyls from the developed seedling (Figure 15.5A).

Meanwhile double coty-nodes are split into two equal halves to obtain single coty-node explants (Figure 15.5B). It is, therefore, important to note that disinfected soybean seeds were germinated on basal MS culture medium supplemented with cytokinins to develop seedlings with thicker hypocotyls that were more suitable for explant preparation (Mangena et al. 2015; Mangena and Mokwala 2019).

15.5 *AGROBACTERIUM TUMEFACIENS* AND CO-CULTIVATION OF EXPLANTS

The *Agrobacterium tumefaciens* is usually cultivated and rejuvenated using yeast extract peptone medium (YEP). Compositions of liquid and solid YEP medium are shown in Table 15.1. The vector system pTF101 was cultured on YEP medium containing 100 mg/L spectinomycin and 50 mg/L kanamycin. Figure 15.6 shows the growth of *Agrobacterium* on solid YEP agar plate (A), YEP liquid medium (B) and (C) infection medium. Figure 15.6 demonstrates cotyledonary explants that were co-cultured with *Agrobacterium* (pTF101.1) for 4 days under controlled tissue culture conditions.

Further information on the composition of the infection medium and co-cultivation medium can be found in Chapter 4. As previously indicated, efficient soybean transformation also depends upon the nutritional supply of carbon, nitrogen and phosphorus required for *A. tumefaciens* growth. The bacterium needs to be rejuvenated before explant infection and co-cultivation stages are performed. Other media used for this purpose may include Luria-Bertani and Yeast Extract Mannitol (YEM) broths (Igbal et al. 2021) (Table 15.1 and Figure 15.7).

15.6 SHOOT INDUCTION

The establishment of adventitious or axillary shoots during soybean transformation can be accomplished on any type of shoot induction medium (MS, B_5, etc.) containing appropriate amounts of micronutrients, macronutrients, iron source ($FeSO_4.7H_2O$), vitamins (B-vitamins), carbon source (preferably sucrose) and PGRs (auxins and cytokinins). Cytokinins such as benzyladenine (BA) and thidiazuron (TDZ) are the most frequently used PGRs because they are not rapidly degraded in culture, and have proved to remain highly stable in *in vitro* tissue culture (Tefera and Wannakrairoj 2006; Siddique et al. 2015) (Figure 15.8).

15.6.1 *In Vitro* Elongation and Rooting

PGRs such as gibberellic acid (GA_3) and auxins [1-naphthaleneacetic acid (NAA), indole-3-butyric acid (IBA) and indole-3-acetic acid (IAA)] are commonly used to

TABLE 15.1
Composition of yeast extract peptone medium

Chemical	Amount (g/L)
Yeast extract	5
Peptone	10
$NaCl_2$	5
Bacto-agar	12 (YEP solid medium)
pH: 7.0	

Source: Paz et al. 2006.

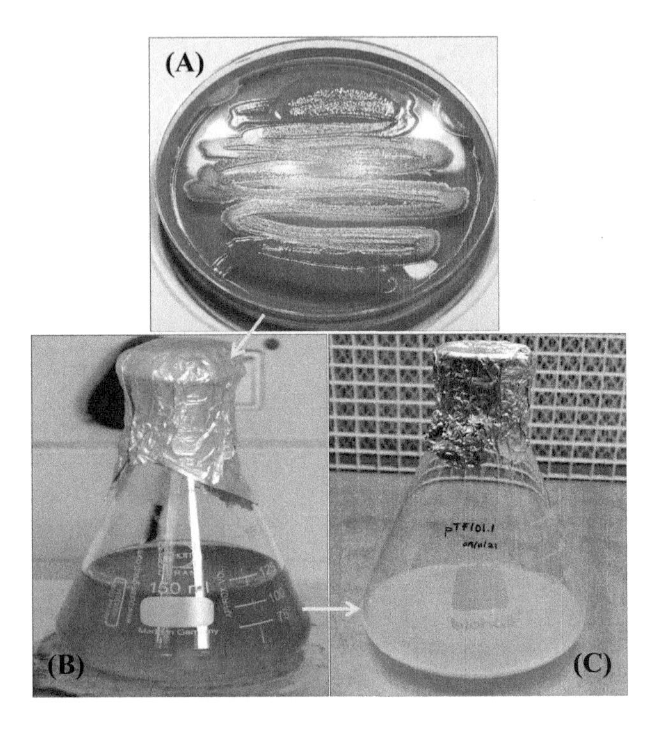

FIGURE 15.6 *Agrobacterium tumefaciens* in liquid culture.

FIGURE 15.7 *Agrobacterium* co-cultivated cotyledonary node explants.

support and direct further development of initiated soybean shoots. Shoot elongation (Figure 15.9A) and rooting (Figure 15.9B) are strongly supported by various auxin and gibberellin concentrations. In general, the minimum concentration of auxins (GA_3, NAA, IAA or IBA) that stimulates elongation or rooting of shoots is determined during culture establishment and varies according to species as well as the goal of the experiment.

FIGURE 15.8 Multiple shoot induction on *Agrobacterium*-co-cultured double coty-node explants.

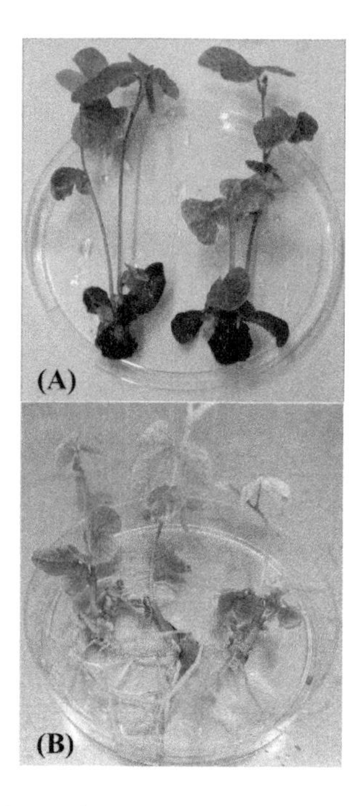

FIGURE 15.9 Elongated and rooted axillary shoots.

15.7 PLANT HARDENING AND ACCLIMATISATION

After the shoots are rooted, they are shifted from heterotrophic (sugar-requiring) to autotrophic (free-living) conditions. Normally, rooted plantlets will be kept in very high humidity and light to prevent dehydration before introducing them to outdoor conditions. Plantlets have to establish new growths and be thoroughly cared for to

FIGURE 15.10 Hardening and acclimatisation of *in vitro* regenerated plantlets.

facilitate acclimatisation. It is important to note that plantlets developed under *in vitro* conditions are unique and different from field-grown plants, especially in terms of the morphology (leaf size, number of cell layers, stomata, cuticle and trichomes) (Figure 15.10).

15.8 GREEN FLUORESCENT PROTEIN (GFP)

Photographs of alternative reporter gene coding for GFP are shown in Figure 15.11 and Figure 15.12. This is also a simple and fast tool for investigating transgene transfer and expression in plants. These photos are courtesy of Joffrey Mejias (Department of Plant Pathology and Microbiology, Iowa State University, United States of America).

15.9 SUMMARY

An improved plant genetic transformation is an absolute necessity to enhance agricultural food production. Crops such as soybean are critically important for achieving the ambitions of feeding the global population, alleviating poverty, minimising food insecurity, growing GDP and providing raw materials for the manufacturing of pharmaceutical products. Current advances in plant regeneration *in vitro* and *Agrobacterium*-mediated genetic transformation have led to routine protocols

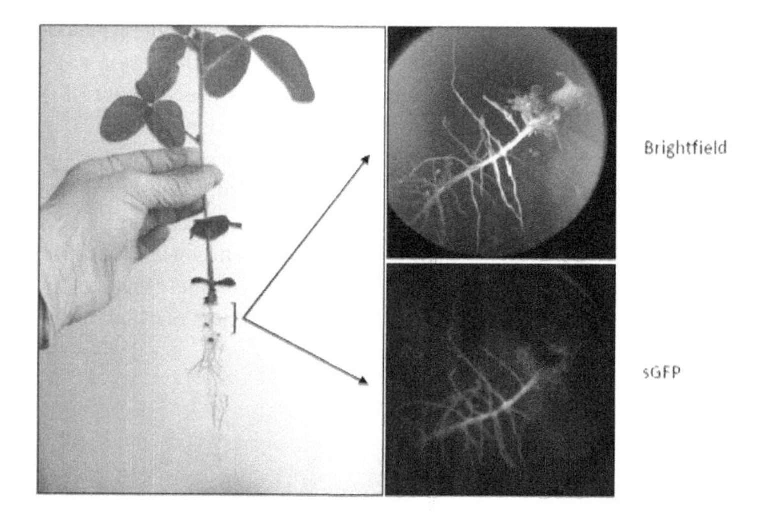

FIGURE 15.11 GFP expressed in soybean root using *Agrobacterium rhizogenes* K599 strain observed using an inflorescent microscope.

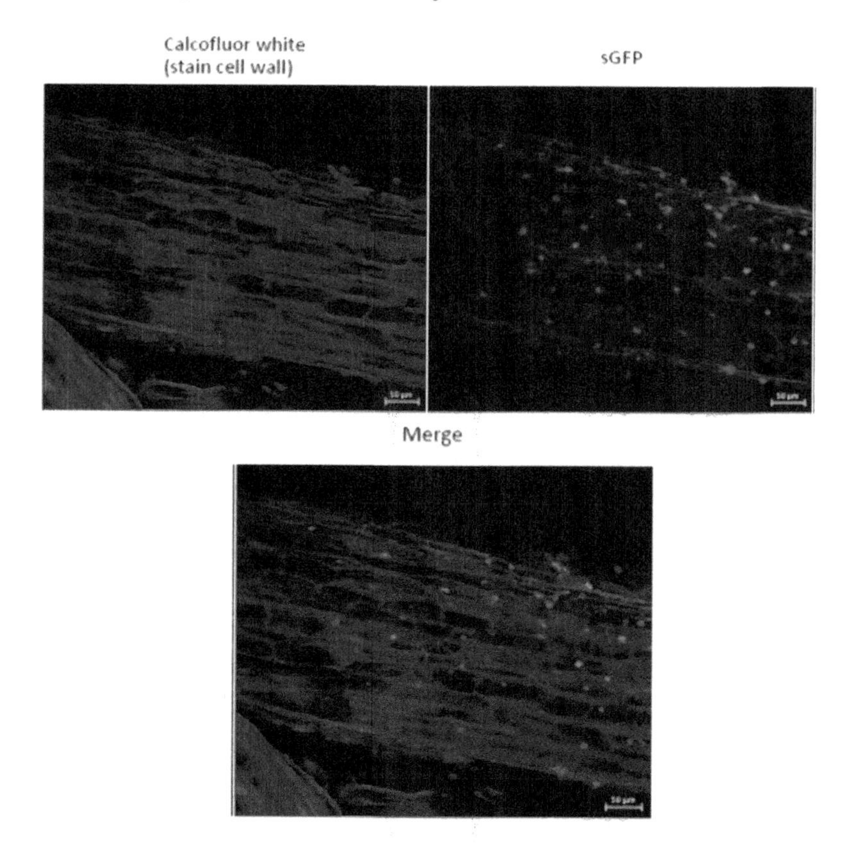

FIGURE 15.12 GFP expressed in soybean root using *A. rhizogenes* K599 strain observed using a confocal microscope.

established for the genetic improvement of monocotyledonous plants and a few dicot species. Here this chapter briefly discussed and illustrated the major stages forming the basis of soybean transformation using *in vitro* plant tissue culture (PTC).

These PTC stages are significant to recognise and identify critical procedures behind genetic transformation. Furthermore, they are essential in the optimisation of genetic transformation in soybean. Plant tissue culture has become a standard propagation procedure for many horticultural crops, and therefore, a key to unravelling genetic recalcitrance in soybean transformation might be found through the application of this technology. It is undisputable that an emerging protocol of plant transformation relies heavily on the efficient regeneration of transformed (genetically engineered) plants from tissue culture.

15.10 ABBREVIATIONS

BA Benzyladenine
GA_3 Gibberellic acid
GDP Gross domestic product
GFP Green fluorescent protein
IAA Indole-3-acetic acid
IBA Indole-3-butyric acid
MS Murashige and Skoog medium
1-NAA 1-naphthalene acetic acid
PGR Plant growth regulator
PTC Plant tissue culture
T-DNA Transfer-DNA
TDZ Thidiazuron
UV Ultraviolet
YEM Yeast extract mannitol
YEP Yeast extract peptone

REFERENCES

Bhajan SK, Begum S, Islam MN, Hoque MI and Sarker RH. (2019). In vitro regeneration and *Agrobacterium*-mediated genetic transformation of local varieties of mungbean (*Vigna radiata* (L.). Wilczek). *Plant Tissue Culture and Biotechnology* 29(1), 81–97.

Carey SB, Payton AC and McDaniel SF. (2015). A method for eliminating bacterial contamination from *in vitro* moss cultures. *Applications in Plant Science* 3(1), 1–5.

Chalfie M, Tu Y, Euskirchen G, Ward WW and Prasher DC. (1994). Green fluorescent protein as a marker for gene expression. *Science* 263(5148), 802–805.

de Oliveira MCP and Costa MGC. (2010). Growth regulators, culture media and antibiotics in the *in vitro* shoot regeneration from mature tissue of Citrus cultivars. *Pesquisa Agropecuaria Brasileira* 45(7), 654–660.

Harper BK and Steward CN. (2000). Patterns of green fluorescent protein expression in transgenic plants. Plant Molecular Biology Reporter 18, 141a–141i.

Iqbal A, Dave N and Kant T. (2021). Comparative analysis of different nutrient media for growth of *Agrobacterium tumefaciens* under small volume cultures. *Journal of Applied Life Sciences International* 24(2), 27–33.

Li X, Cong Y, Liu Y, Wang T, Shai Q, Chen N, Gai J and Li Y. (2017). Optimisation of *Agrobacterium*-mediated transformation in soybean. *Frontiers in Plant Science* 8(246), 1–15.

Mangena P. (2018). Water stress: Morphological and anatomical changes in soybean (*Glycine max* L.) plants. In Andjelkovic V. (eds), *Plant, Abiotic Stress and Responses to Climate Change*. Intech Open, London. pp. 9–31.

Mangena P. (2019). A simplified in-planta genetic transformation in soybean. *Research Journal of Biotechnology* 14(9), 117–125.

Mangena P. (2021). Effect of *Agrobacterium* co-cultivation stage on explant response for subsequent genetic transformation in soybean (*Glycine max* (L.) Merr.). *Plant Science Today* 8(4), 905–911.

Mangena P and Mokwala PW. (2019). The influence of seed viability on the germination and *in vitro* multiple shoot regeneration of soybean (*Glycine max* L.). *Agriculture* 9(35), 1–12.

Mangena P, Mokwala PW and Nikolova RV. (2015). *In vitro* multiple shoot induction in soybean. *International Journal of Agriculture and Biology* 17, 838–842.

Opabode JT. (2006). *Agrobacterium*-mediated transformation of plants: Emerging factors that influence efficiency. *Biotechnology and Molecular Biology Review* 1(1), 12–20.

Paz M, Martinez JC, Kalvig A, Fonger T and Wang K. (2006). Improved cotyledonary node method using an alternative explant derived from mature seed for efficient *Agrobacterium*-mediated soybean transformation. *Plant Cell Reports* 25, 206–213.

Paz MM, Shou H, Guo Z, Zhang Z, Banerjee A and Wang K. (2004). Assessment of conditions affecting *Agrobacterium*-mediated soybean transformation using the cotyledonary node explant. *Euphytica* 136, 167–179.

Siddique I, Bukhari NAW, Parveen K and Siddiqui I. (2015). Influence of plant growth regulators on *in vitro* shoot multiplication and plantlet formation in *Cassia angustifolia* Vahl. *Brazilian Archives of Biology and Technology* 58(5), 686–691.

Tefera W and Wannakrairoj S. (2006). Synergistic effects of some plant growth regulator on *in vitro* shoot proliferation of korarima (*Aframomum corrorima* (Braun) Jansen). *African Journal of Biotechnology* 5(10), 1894–1901.

Zhang F, Chen C, Ge H, Liu J, Luo Y, Liu K, Chen L, Xu K, Zhang Y, Tan G and Li C. (2014). Efficient soybean regeneration and *Agrobacterium*-mediated transformation using a whole cotyledonary node as an explant. *Biotechnology and Applied Biochemistry* 61(5), 620–625.

Glossary

Abiotic stress: the negative impact of non-living factors on the living organisms in a specific environment.

Abscission: detachment of plant organs from the mother plant.

Adenosine triphosphate (ATP): the major carrier of energy (chemical energy) in the cell, which is normally hydrolysed to form adenosine diphosphate or adenosine monophosphate.

Adventitious: cell proliferation from an unusual origin, such as the development of shoots or roots.

Antibiotics: chemical substances, often derived from various microorganisms, that have the capacity to inhibit microbial growth or destroy them.

Aseptic culture: conditions, usually in tissue culture that are free from contaminating microorganisms.

Autoclave: a pressurised machine designed to heat aqueous solutions above their boiling points at normal atmospheric pressure to achieve sterilisation. This device is also used to sterilise glassware and instruments.

Beta-glucuronidase (*Gus*): a reporter gene allowing a spatial and temporal expression of the gene present in tissues and organs of transgenic plants. The gene is used for evaluating transient and stable transformation in plants.

Binary fission: type of asexual reproduction where a parent cell divides, resulting in two identical daughter cells, taking place in bacteria and organisms like amoeba.

Binary vector: a pair of plasmid DNA molecules consisting of the T-DNA and virulence genes serving as a tool of genetic transformation in higher plants.

Biotic stress: the negative impact of living factors on other living organisms in a specific environment.

Biovar: group of microorganisms, usually bacteria, that possess identical genetic materials but differ in biochemical or physiological characteristics.

Bt toxins: toxic substances produced by bacterium *Bacillus thuringiensis* that are fatal to certain herbivorous insects.

Callus: unorganised mass of cells. Callus can be transferred onto a different culture media to regenerate plants.

Cell culture: an *in vitro* growth of cells isolated from multicellular or unicellular organism.

Cell: refers to a structural, functional and biological unit of an organism.

Chimeras: plant or plant part growing from two or more genetically different types of cells.

Chromosome: a threadlike structure found in nucleic acids found in the nucleus of living organisms

Circular DNA: DNA that forms a closed loop and has no end.

Clone: a group of cells or organisms produced asexually from one ancestor and they are genetically identical.

Cotyledon: an embryogenic leaf developing from a germinating seed.

Cryopreservation: ultra-low temperature storage of cells, tissues, embryos or seeds.

Culture condition: artificial growth conditions designed for culturing of cells, tissue or protoplasts, usually under controlled conditions of light, temperature and humidity.

***De novo*:** of roots or shoots, is a process in which adventitious roots or shoots regenerate from detached or wounded plant tissues.

Deoxyribonucleic acid (DNA): a self-replicating material constituting the chromosomes of all living cells.

Dicotyledonous plant: member of the flowering plants (Angiosperms) that has a pair of leaves or cotyledons.

Donor plant: plant (or plant part) used in tissue culture as a source of an explant.

Embryogenesis: the formation of an embryo from a zygote or mass of undifferentiated callus cells.

Epigenetic: changes in the DNA molecule caused by methylation.

***Ex vitro*:** an environment outside the artificial tissue culture conditions, usually in soil or potting mixture.

Explant: an excised piece of tissue or organ taken from the donor plant, used to initiate a culture.

Flow cytometry: technique used in the detection and measurement of cells' characteristics such as number, size and nucleic acid content.

Gene pool: a stock of different genes in an interbreeding population.

Gene: unit of hereditary or distinct sequence of nucleotide forming part of the chromosome.

Genetic transformation: process of transferring and incorporating foreign genetic materials into a host genome.

Genetically modified organism (GMO): organism whose genome has been altered using a genetic engineering technique in order to favour the expression of certain desirable genes.

Genotype: the genetic constitution of an individual organism.

Germplasm: collection of genes, particularly in the form of seeds, plants or plant parts useful in crop breeding.

Gibberellins: a large group of chemically related plant hormones synthesised by a branch of the terpenoid pathway and associated with the promotion of stem growth, seed germination and many other functions.

Habituation: refers to an independent or uncontrolled division and growth of plant cells without plant growth regulators (cytokinins).

Horizontal gene transfer: transfer or movement of genetic materials between unicellular and multicellular organisms without the involvement of sexual reproduction.

Hypersensitive reaction (HR): a common plant defence following exposure of plants to biotic and/or abiotic stress.

***In vitro* regeneration:** a plant tissue culture system in which cells and tissues undergo cell division and differentiation, forming new organs and a whole new plant.

***In vitro*:** a process of culturing plant cells, tissues or organ on artificial media, in an aseptic and controlled environment.

***In vivo*:** conditions outside the tissue culture environment.

Lipopolysaccharide: large cell wall molecule consisting of lipids and sugars joined by chemical bonds.

Meristem: a region of plant tissue found at the growing tips of roots and shoots.

Microprojectile bombardment: technique employing high-velocity metal particles to deliver foreign DNA molecules into plant cells.

Micropropagation: production of a large number of clones using plant tissue culture.

Monocotyledonous plant: member of the flowering plants (Angiosperms) that has a single leaf or cotyledon.

Murein lipoprotein: a major outer membrane lipoprotein found in Gram-negative bacteria.

Mutant: individual plant with altered characteristics, especially the changes in folia, stem, flower or fruits.

Mutation breeding: a process of exposing plant materials (seeds) to chemicals, radiation or enzymatic treatments in order to generate mutants with desirable traits.

Oncogene: a mutated gene with the ability to transform a cell into a tumour cell. Such a gene can be transmissible, leading to the development of cancers in a progeny.

Opines: low molecular weight carbohydrate derivatives found in plant crown gall tumours or hairy root tumours that serve as a nutrient source for the Agrobacterium.

Organogenesis: development of organs like roots, stems, leaves and flowers either directly from an explant or indirectly through a callus culture.

Pathogen: bacteria, fungi, virus or other microorganism that can cause a disease.

Peptidoglycan: major structural polymer found in bacterial cell walls, consisting of glycan chains of repeating N-acetylglucosamine and N-acetylmuramic acid residues cross-linked via peptide side chains.

Phenotype: a set of observable characteristics of an individual resulting from the interactions involving the genotype and the environment.

Phospholipids: a group of polar lipids consisting of two fatty acids, a glycerol unit and a phosphate group which is esterified to an alcohol residue.

Photoperiod: period of time each day during which an organism receives illumination, also referred to as a day length.

Phytohormone: chemicals produced by plants that regulate their growth, development, reproductive processes, longevity and even death.

Pili: short filamentous projection on a bacterial cell wall used for motility.

Plant growth regulator (PGR): chemicals used for modifying the growth of plants, especially in plant tissue culture such as shoot proliferation, blooming or fruiting and even ripening.

Plant tissue culture (PTC): the process of culturing plant seeds, organs, explants, tissues, cells or protoplasts on a chemically defined synthetic nutrient media.

Plant varieties: a more precisely defined group of plants, selected from within a species and containing a common set of phenotypic and genotypic characteristics.

Plasma membrane: a fluid mosaic structure composed of a lipid bilayer (phospholipids or glycolipids) and embedded proteins that work together to confer selective permeability on the structure.

Polyethylene glycol (PEG): a synthetic, hydrophilic biocompatible polyether used chiefly as solvents or waxes.

Polymerase chain reaction (PCR): a method of making multiple copies of DNA involving repeated reactions using an RNA/DNA polymerase enzyme.

Programmed cell death (PCD): death of a cell as a result of apoptosis or autophagy.

Protease inhibitors (PIs): a group of proteins that regulate proteolytic activity of protease enzymes. PIs are widely distributed in living organisms like animals, fungi, bacteria and plants.

Protoplast: a living part of plant cell or bacterial cell whose cell wall has been removed.

Qualitative trait loci: chromosomal region that contains genes that control a quantitative trait.

Reactive oxygen species (ROS): highly reactive radicals or species formed from oxygen that are capable of independent existence with one or more unpaired electrons.

Recalcitrant seeds: seeds that cannot survive desiccation or undergo no maturation drying as the final stage of development.

Rhizosphere: region of soil in the vicinity of plant roots in which the chemistry and microbiology are influenced by root growth.

Ribonucleic acid (RNA): nucleic acid present in all living organisms, functioning as a messenger carrying instructions from the DNA in controlling the synthesis of proteins.

Selectable markers: gene introduced in a host genome that confers a trait suitable for artificial selection.

Somatic embryogenesis: an artificial tissue culture-based process in which a plant or an embryo is established from a single somatic cell.

Southern blotting: a technique for identifying a specific sequence of DNA in which fragments separated on a gel are transferred directly to a second medium on which hybridisation assays are carried out.

T-complex: multicomponent protein complex used to deliver macromolecules directly into their eukaryotic host cell to promote infection by pathogens.

Tetraploids: a cell or nucleus containing four homologous sets of chromosomes.

Ti-plasmid: a tumour-inducing genetic structure found in pathogenic species of *Agrobacterium* essential for the bacterium to cause crown gall disease in plants.

Tissue senescence: process of aging, often leading to abscission of plant organs.

Totipotency: ability of a cell to divide and produce all differentiated cells in an organism that can autonomously develop into a whole plant.

Transferred DNA (T-DNA): genetic structure of a plasmid DNA containing virulence genes which control the process of infection of plants and genes expressing specific compounds, opines, which are used by *Agrobacterium* as a carbon source.

Transgenic plants: plants whose DNA is modified through genetic engineering.

Ultraviolet (UV): a form of electromagnetic radiation with wavelength from 10 nm to 400 nm, which is shorter than visible light but longer than X-ray.

Virulence genes: genetic materials that regulate and enable microbial pathogens to colonise and infect resistant hosts.

Yeast extract peptone (YEP): a complete medium used for the growth of microorganisms.

Index

F

G

H

I

K

L

M

N

O

P

Q

R

S

T

V

Y

Z

9781032250380